# INCREDIBLE LEGO® TECHNIC

# INCREDIBLE LEGO® TECHNIC
## CARS, TRUCKS, ROBOTS & MORE!

## Paweł "Sariel" Kmieć

NO STARCH PRESS
SAN FRANCISCO

**Incredible LEGO® Technic.** Copyright © 2015 by Paweł "Sariel" Kmieć.

Printed in China
First printing

18 17 16 15 14    1 2 3 4 5 6 7 8 9
ISBN-10: 1-59327-587-0
ISBN-13: 978-1-59327-587-7

Publisher: William Pollock
Production Editor: Serena Yang
Cover Design: Beth Middleworth
Interior Design: Ryan Byarlay and Beth Middleworth
Developmental Editor: Tyler Ortman
Technical Adviser: Eric "Blakbird" Albrecht
Copyeditor: Kim Wimpsett
Compositor: Serena Yang
Proofreaders: Lisa Devoto Farrell and Alison Law

**Cover Models**
Front: Koenigsegg CCX, Jurgen Krooshoop
Back: Land Rover Defender 110, Fernando Benavides de Carlos
       Stan Tug 4011 SL Gabon, Edwin Korstanje
       Tachikoma, Peer Kreuger
       Caterpillar D9T, Ignat Khliebnikov

For information on distribution, translations, or bulk sales, please contact No Starch Press, Inc. directly:
No Starch Press, Inc.
245 8th Street, San Francisco, CA 94103
phone: 415.863.9900; info@nostarch.com;
www.nostarch.com

Library of Congress Cataloging-in-Publication Data
A catalog record of this book is available from the Library of Congress.

Photo credit: Weronika Łojek

## ABOUT THE AUTHOR

Paweł "Sariel" Kmieć is a LEGO Technic enthusiast based in Warsaw, Poland, and the author of the popular *Unofficial LEGO Technic Builder's Guide* (No Starch Press). A prolific builder, he is known mostly for his models of tanks and trucks, but his pursuit of new ideas has led him to dabble in aircraft, watercraft, and even monorail creations. The LEGO Group used his help in developing its Power Functions system and sends him Technic sets for review. He was a 2012 LEGO Ambassador for Poland and one of just three people in the world—together with Eric Albrecht and Fernando Correia—enrolled as a guest blogger for the LEGO Technic website. He's also a successful YouTuber with over 20 million views, which may have to do with the hamsters that appear in his videos when you least expect them.

## ABOUT THE TECHNICAL ADVISER

Eric "Blakbird" Albrecht is an aerospace engineer living in the northwest United States. He maintains the globally known *Technicopedia* (*http://www.technicopedia.com/*), in which he documents the history of every Technic model from 1977 on, all of which are on display on his shelves. Eric is also an avid user of LEGO CAD tools and has generated over 1,000 photo-realistic renders of official models and MOCs (My Own Creations). He has created dozens of sets of instructions for some of the best Technic MOCs from around the world, averaging 1,500 parts each. He was the technical reviewer for *The Unofficial LEGO Technic Builder's Guide* and illustrated *The Unofficial LEGO Builder's Guide, 2nd Edition* (both from No Starch Press).

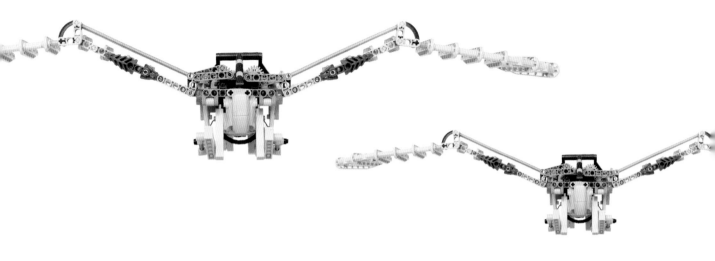

# CONTENTS

## FOREWORD

LEGO bricks are an excellent medium for expression that let you create amazing models and even art, but LEGO tends to be mostly static. The LEGO Technic system, first introduced in 1977, turns that assumption on its head; it brings kinetic sculptures, mechanical elements, and dramatic movements into the otherwise fixed and immobile world of LEGO.

Like any successful LEGO product, the Technic system was embraced by a community of amazing fan builders not content to merely re-create the LEGO Group's official designs. Today, with decades' worth of pieces to work with and freed from any design guidelines, these builders create Technic models that shock and amaze, even blurring the line between LEGO and reality.

When I heard Paweł was working on his first book, *The Unofficial LEGO Technic Builder's Guide*, I immediately wished for a photo book showing off his finished creations, as I was familiar with all his amazing Technic models built over the years. With his second book, Paweł succeeds in that premise, but he also celebrates the community's work by inviting other world-class Technic builders to join him, showcasing a range of amazing styles and subjects.

I'm grateful to see so many models that I've highlighted myself at TechnicBRICKs.com over the last few years presented together in this gorgeous book. The high-quality photos and renderings are stunning and give even familiar creations a whole new perspective. In short, I hope you find, as I did, that this is the LEGO Technic book you always wanted to have on your bookshelf.

Fernando "Conchas" Correia
Editor in Chief, TechnicBRICKs.com
Lisbon, 2014

## PREFACE

The book you are about to read is a result of a collective effort rather than a single author's work. My job was to select and gather the models showcased in this book and to present them in the best possible manner. But I couldn't have done that without the many awesome builders who agreed to join this project.

Some of them have even rebuilt their models just for this book, or have taken dozens of new photos, and I thank them for it. Their support says a great deal about the community of AFOLs (adult fans of LEGO), as we call ourselves. It demonstrates how this community, made up of many individuals striving to better themselves, is not about rivalry but respect and support. Even when two members of this community happen to build models of the same machine, the inevitable comparison is geared toward mutual appreciation rather than competition. This is what makes this community special, and this community is what made this book possible.

One question was essential to the making of this book: what is a Technic model? In the end, I tried to include any model that goes to significant lengths to re-create the workings of a real machine. While this definition is broad and somewhat ambiguous, it allowed me and Eric Albrecht, the technical adviser, to make selections that serve this book's goal: to demonstrate that LEGO Technic can be much more than just the sets you find in a store; it has great potential to be combined with pretty much any element from the vast family of LEGO products. In this book you will find models whose creators worked long and hard to follow the style of modern Technic sets as closely as possible, but you'll also find models that look utterly non-Technic, which remind us that even more than 50 years after the invention of the LEGO brick, every LEGO element remains a component in a single, larger system.

For me personally, this book is a dream come true. I've been looking for an opportunity to help other builders showcase their creations, and this book provides it. It also allows Eric to use his extraordinary talent in 3D rendering to present the models we have chosen in the best way possible.

The selection of models we have made is, by necessity, subjective and limited. Some of the models we wanted to include simply did not have photographs suitable for print; some could not be shown for other reasons. My only regret is that there are countless new amazing models presented every day, and that the moment this book is finished I will start seeing models I wish could be a part of it. However, we did the best we could so that our selection serves our primary goal: to show you the possibilities of the LEGO system and to encourage you to explore them. The models we have chosen aren't here just to dazzle—they are here to inspire.

I hope they will do just that for you.

Sariel

## ACKNOWLEDGMENTS

This book could have never happened without Eric "Blakbird" Albrecht, whose invaluable expertise was essential in shaping its direction and whose skills helped a great deal in presenting its content.

And the book certainly wouldn't be possible without the many outstanding, talented LEGO Technic enthusiasts who allowed me to share their amazing work, often going to great lengths to create high-quality visuals, and who were always kind, patient, and understanding, even when harassed with countless emails and deadlines. It was an amazing, unique experience to turn to so many people for help and receive support every time.

Of all the builders I asked to contribute to this project, not a single one said "no." That's a wonderful testament to how respectful and close-knit this community is, and I want to express my gratitude to all who helped. Of course, due to practical reasons and circumstance beyond our control, not every model that deserves to be showcased in print appears in these pages. But I still hope that it serves the reader well as a representative sample of the best work being done with LEGO Technic.

Special thanks go to Tyler Ortman, Serena Yang, Ryan Byarlay, and the rest of the No Starch Press crew for helping to make this book better than I could have ever planned.

Also, thanks to whoever invented coffee for helping me finish this book on time.

Last but not least, many thanks to all who enjoyed and supported my previous book, because it's you who gave me the chance and incentive to continue writing.

# AGRICULTURAL EQUIPMENT

# DT-75 TRACTOR

*Dmac (2009)*

## ABOUT THE MODEL

Combining electric and pneumatic elements, this model is well suited for a variety of tasks, from agricultural jobs to work as a bulldozer. Equipped with an internal pneumatic compressor and a remotely controlled pneumatic switch, the model is relatively small and tightly packed with functions, while looking like anything but Technic. It comes with a removable blade in the front and a working power take-off (PTO) in the back, which could power external attachments such as a plow or a ripper.

## CHALLENGES

Finding red pieces was the biggest challenge of this model. Choosing red required using the older type of pneumatic cylinders, and it forced me to search far and wide for many rare pieces, including wagon wheels from a LEGO Castle set. These wheels, used as the road wheels in the tracks of my model, don't officially exist in red. The ones I bought were produced unofficially, probably to test molds. It's said that LEGO uses red plastic in tests because the color makes it easier to spot flaws.

## THE ORIGINAL

The DT-75 tracked tractor was a workhorse in the former Republics of the Soviet Union. With a wide array of additional equipment, this versatile machine was the only choice for a lot of farmers for many years. It is simple, bordering on crude, but dependable even in severe weather. Its simple mechanisms mean the most basic tools are all you need when things break down.

## SPECIFICATIONS

| | |
|---|---|
| LENGTH | **14.2"** |
| WIDTH | **6.3"** |
| HEIGHT | **7.9"** |
| PIECES | **~2,000** |

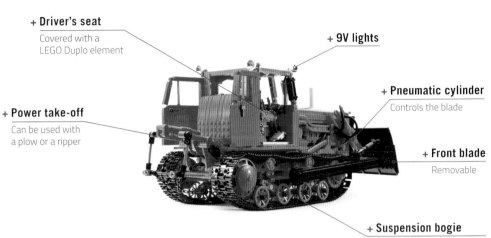

**+ Driver's seat**
Covered with a
LEGO Duplo element

**+ 9V lights**

**+ Pneumatic cylinder**
Controls the blade

**+ Power take-off**
Can be used with
a plow or a ripper

**+ Front blade**
Removable

**+ Suspension bogie**
With four wagon wheels
and a single shock absorber

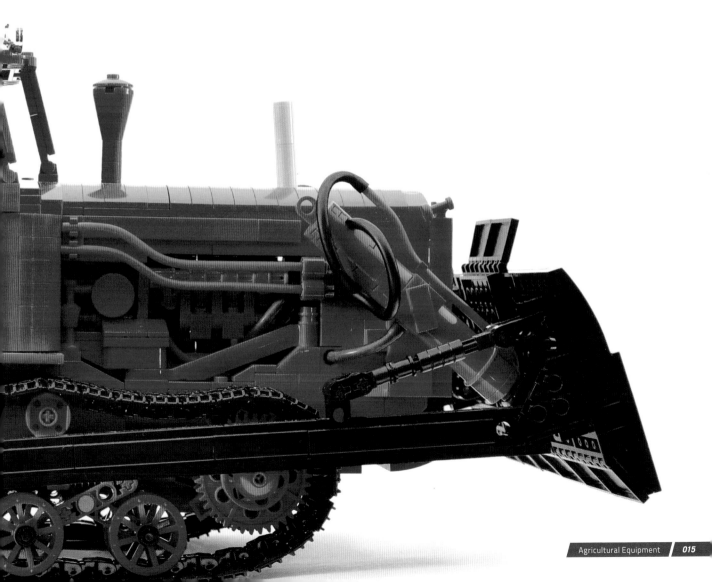

You start the DT-75 tractor via a recoil starter, just like a lawn mower. The starter activates a small combustion engine, which then starts the huge diesel engine that powers the tracks.

**+ FUN FACT**
LEGO DUPLO and Scala
elements are used to
imitate the original seat
covers.

# HOLMER TERRA DOS T3

*Eric Trax (2013)*

*The Holmer Terra Dos T3 extracts enough sugar beets to fill a dump truck in just 10 minutes.*

## SPECIFICATIONS

| | |
|---|---|
| LENGTH | **32.6"** |
| WIDTH | **8.7"** |
| HEIGHT | **14.2"** |
| PIECES | **~6,000** |

## ABOUT THE MODEL

This model of a Holmer sugar beet harvester balances looks and functionality. The harvester is operated by 11 motors that are all concealed, and it features all-wheel drive, rear-axle steering, articulated front-axle steering, working conveyor belts with adjustable elevation of the unloading belt, rotating harvesting heads, and working lights. Its impressive functionality can be topped only by its enormous size and meticulously re-created exterior.

## CHALLENGES

This massive model weighs 13 lb (5.9 kg), which was a serious challenge. To carry that much weight, the body was built around a truss body frame and equipped with wheels mounted on roller bearings. The rear wheels handle nearly 4.5 lb (2 kg) each, which normally would deform the tires and increase the rolling resistance drastically. To prevent this, the wheels are equipped with double tires: Smaller ones are inserted inside the external ones.

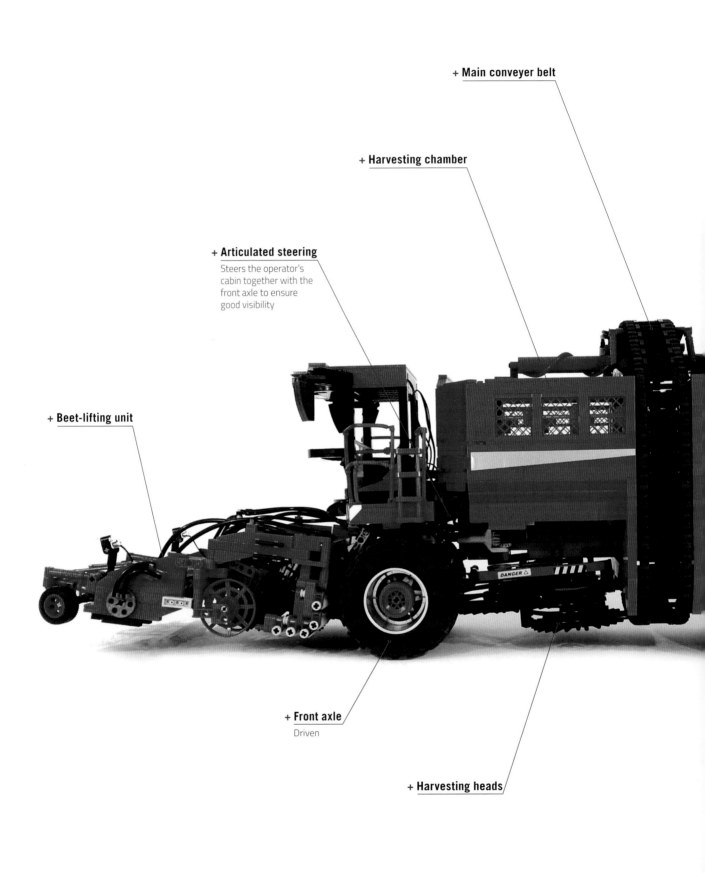

+ **Main conveyer belt**

+ **Harvesting chamber**

+ **Articulated steering**
Steers the operator's
cabin together with the
front axle to ensure
good visibility

+ **Beet-lifting unit**

+ **Front axle**
Driven

+ **Harvesting heads**

DANGER ⚠

**+ Unloading conveyer belt**
Can be elevated

**+ Rear axle**
Driven and
steered

### THE ORIGINAL

The German Holmer company has 40 years of
experience in building sugar beet harvesters.
The Terra Dos T3 is the third version of the
flagship Holmer machine, and it was designed
to extract and preprocess massive amounts of
sugar beets while allowing just a single operator
to control the harvester and to easily maneuver
this 40-foot (12-meter) long colossus in rough
terrain.

# LANZ BULLDOG HOT BULB TRACTOR

*Nico71 (2012)*

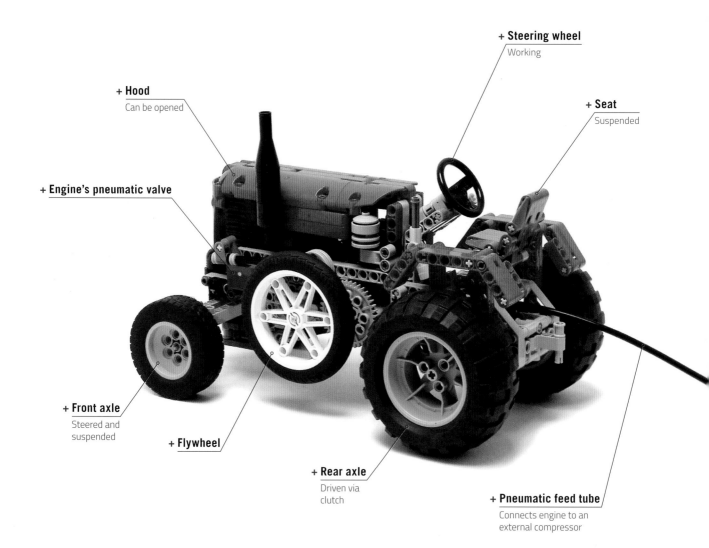

**+ Steering wheel**
Working

**+ Hood**
Can be opened

**+ Seat**
Suspended

**+ Engine's pneumatic valve**

**+ Front axle**
Steered and
suspended

**+ Flywheel**

**+ Rear axle**
Driven via
clutch

**+ Pneumatic feed tube**
Connects engine to an
external compressor

*Even though Lanz ceased operations more than half a century ago, tractors are still referred to as "bulldogs" in many areas of Germany.*

## SPECIFICATIONS

LENGTH **11.3"**

WIDTH **6"**

HEIGHT **7.5"**

PIECES **~600**

## ABOUT THE MODEL

This model of a classic Lanz tractor features a pneumatic engine with an adjustable ignition point, a working steering wheel, a suspended front axle, and a suspended seat. It was built to demonstrate the working principle of hot bulb engines, which were invented in the late 19th century and remained in use until the 1950s. Designed as a simpler alternative to diesel engines, these engines used bulb-shaped ignition chambers, which gave them their name.

The model mimics the behavior of hot bulb engines with a single-cylinder pneumatic engine with a flywheel on the tractor's side and an adjustable ignition point. Adjustment is possible with a worm gear that allows the operator to easily find the optimum starting and running points in the engine's cycle.

This model's engine is powered by a compressor outside the model and can run in idle mode thanks to a simple clutch in the rear axle. The tractor's bodywork includes a large number of details typical for real hot bulb tractors, including a radiator with a rotating fan and the bonnet, which can be opened.

## THE ORIGINAL

The Bulldog was produced by Lanz, a German manufacturer, with more than 220,000 tractors sold between 1921 and 1960. The company was later purchased by John Deere, and its renowned hot bulb tractors were discontinued. The simple and reliable Lanz tractors were easy to maintain and could run even on waste oils, which made them popular with farmers and also prompted a number of copycat designs from competitors.

# URSUS C-360-3P

*Eric Trax (2014)*

## SPECIFICATIONS

| | |
|---|---|
| LENGTH | **11.4"** |
| WIDTH | **7.7"** |
| HEIGHT | **9"** |
| PIECES | **~900** |

## ABOUT THE MODEL

This ultrarealistic model of the Ursus tractor doesn't look like Technic, yet it can be driven with a remote control and comes with realistic pushrod-operated steering, a motorized power take-off (PTO), and arms for attaching accessories that can be elevated. It took three months of work to conceal all the Technic mechanisms inside the Model Team–styled body; in addition, to get wheels big enough for the scale, the Power Puller tires had to be put on custom rims and squeezed for increased diameter.

## CHALLENGES

The model's body is rather small, and it was difficult to fit so many elements inside it, especially with the battery taking up half the space inside the hood. In the end, one of two PF IR receivers is sitting inside the cabin, and so is the steering motor.

+ **Power take-off**
Motorized

+ **Accessory attachment arm**
Elevated

+ **Toolbox**
Can be opened

+ **PF IR receiver**

+ **Power Puller tires**
Squeezed with
custom-built rims
for more accurate
shape

+ **Pushrod**
Operates the
steering system

*The original Ursus
tractor is known for
its noisy transmis-
sion. This is because
rather than using
helical gear wheels,
it's using spur gear
wheels, which are
exactly like LEGO
gear wheels.*

## THE ORIGINAL

The Ursus C-360 3P is a medium-sized tractor,
manufactured in Poland between 1981 and
1995. With easy maintenance and a simple
straight-three engine developing 47.5 hp, the
Ursus is a common sight in the Polish country-
side. It's compatible with a huge array of equipment, and there was even a variant converted
into a small excavator.

# 2

# AIRCRAFT

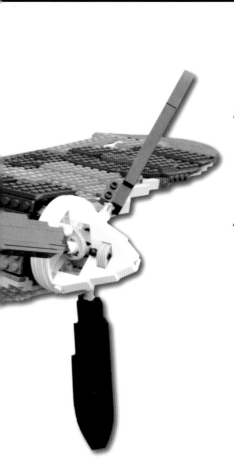

# THE BAT

*Sariel (2013)*

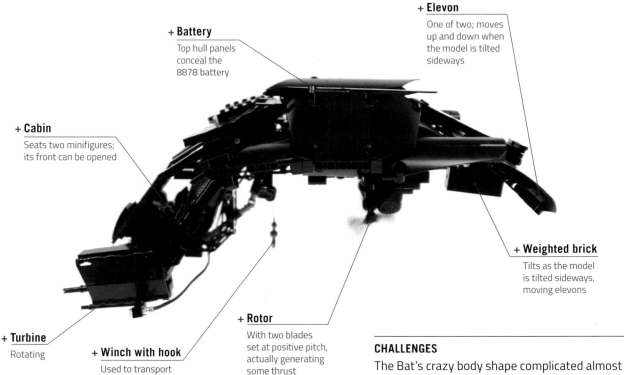

**+ Battery**
Top hull panels conceal the 8878 battery

**+ Elevon**
One of two; moves up and down when the model is tilted sideways

**+ Cabin**
Seats two minifigures; its front can be opened

**+ Weighted brick**
Tilts as the model is tilted sideways, moving elevons

**+ Turbine**
Rotating

**+ Winch with hook**
Used to transport a bomb in the movie finale

**+ Rotor**
With two blades set at positive pitch, actually generating some thrust

### ABOUT THE MODEL

This model of the Bat flyer from *The Dark Knight Rises* features 2 front turbines and 2 belly rotors powered by a single motor, self-balancing elevons, a cabin that seats 2 mini-figures, and 10 LEGO LEDs.

It can "fly" around, because it is attached by a monofilament to a simple bogie that drives along a pipe. Devoid of any actual chassis, it has good looks with plenty of mechanical elements crammed in. It even includes the winch that proved crucial in the movie's finale!

### CHALLENGES

The Bat's crazy body shape complicated almost everything. The large part of the model was built starting from the top, and it had to be kept suspended for most of its construction.

The body needed to include a battery, a motor, several yards of wires, and a complex drivetrain that connects front turbines and belly rotors, all set at very specific angles. Finally, the model had to be properly balanced, which required adding a weighted brick in the back, making self-balancing elevons possible.

### THE ORIGINAL

This heavily armed two-seater hovers (thanks to two belly rotors) and plays a major role in the climax of Nolan's trilogy. The movie crew designed it to match the aesthetic of other vehicles, such as the Tumbler on page 76, drawing inspiration from the Apache helicopter, the Harrier jet, and the V-22 Osprey transporter.

LENGTH **15.7"**

WIDTH **10.7"**

HEIGHT **6.9"**

PIECES **~600**

# LOCKHEED SR-71 BLACKBIRD

*Sariel (2013)*

## SPECIFICATIONS

| | |
|---|---|
| LENGTH | **28.7"** |
| WIDTH | **14.2"** |
| HEIGHT | **3.9"** |
| PIECES | **~900** |

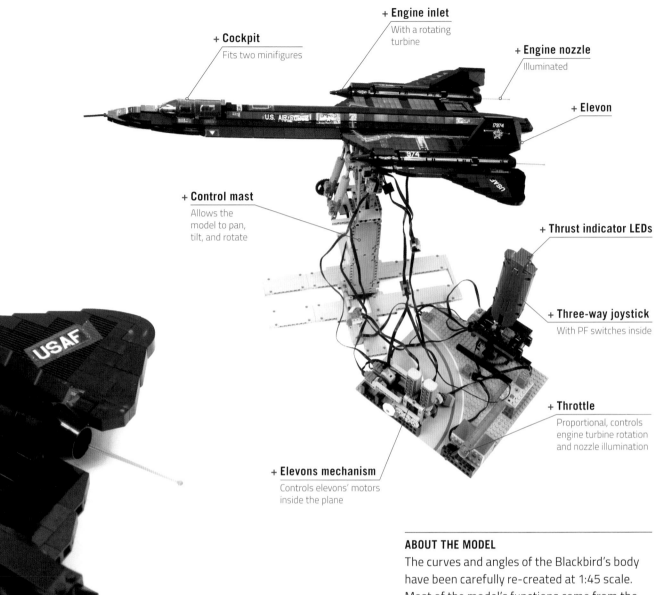

+ **Cockpit**
Fits two minifigures

+ **Engine inlet**
With a rotating turbine

+ **Engine nozzle**
Illuminated

+ **Elevon**

+ **Control mast**
Allows the model to pan, tilt, and rotate

+ **Thrust indicator LEDs**

+ **Three-way joystick**
With PF switches inside

+ **Throttle**
Proportional, controls engine turbine rotation and nozzle illumination

+ **Elevons mechanism**
Controls elevons' motors inside the plane

## ABOUT THE MODEL

The curves and angles of the Blackbird's body have been carefully re-created at 1:45 scale. Most of the model's functions come from the control mast that the plane is mounted on. The mast is connected to a three-way joystick with PF switches inside, which allows the model to pan, tilt, and rotate. The plane has moving elevons, rotating engine turbines, and illuminated jet nozzles, as well as two minifigure pilots seated in the cockpit, and it's covered in an array of custom-made stickers.

## CHALLENGES

The wings! The wings' outer shape had to be painstakingly approximated with small pieces, and the corrugated surfaces on their centers had to be re-created. In addition, the wings had to be sturdy enough to stay together while supporting heavy engines and thick enough to house the driveshafts that connect the engines with the main hull. Finally, they needed the motorized elevons integrated into them.

## THE ORIGINAL

Conceived as the ultimate spy plane, the SR-71 Blackbird entered service in 1966 and remained active until the mid-1990s. To this day it holds a number of records, including the speed record for a manned air-breathing aircraft that was set in 1976. With its unique engine design, which involved constantly engaged afterburners, the plane could operate at speeds up to Mach 3.3 and altitudes up to 85,000 feet (26,000 meters.)

The Blackbird's structure was designed so that parts of the fuselage fit loosely when on the ground but then expanded into their final shape when in the air. Because of this, the plane leaked fuel when on the ground and needed to be refueled in the air after takeoff. To make this practice safe, the Blackbird used specially developed fuel that ignited only at very high temperatures.

### + FUN FACT

*At its top speed, the plane became so hot that it grew several inches longer, which is why a large part of its wings' surface was corrugated and why 85 percent of its structure was pure titanium.*

It is estimated that some 4,000 missiles were fired at Blackbirds in the course of 3,551 missions. None of them hit—the Blackbirds were simply too fast. Their standard evasive maneuver was to outfly any threats.

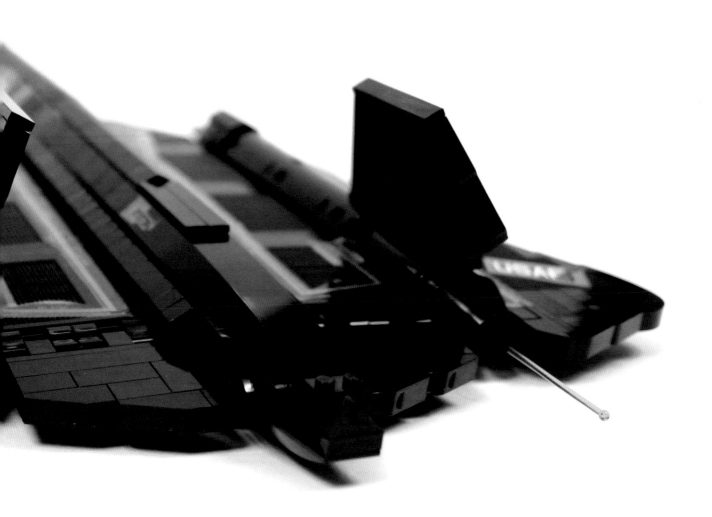

# SA-2 SAMSON
# BATTLE HELICOPTER

*Barman76 (2010)*

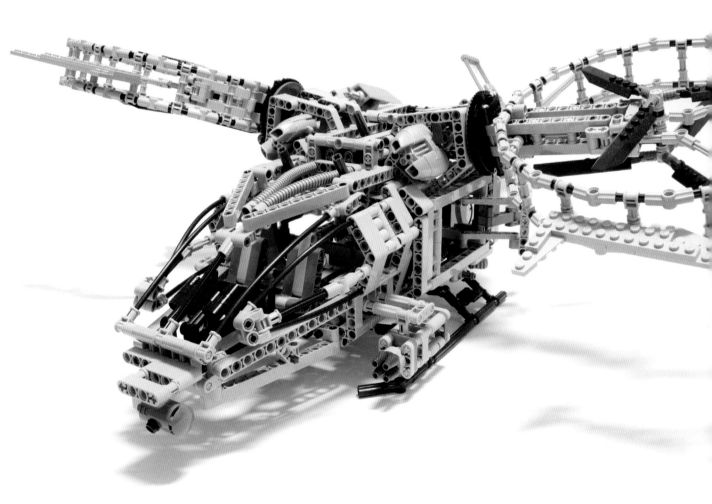

## SPECIFICATIONS

| | |
|---|---|
| LENGTH | **23.4"** |
| WIDTH | **25.4"** |
| HEIGHT | **8.6"** |
| PIECES | **1,823** |

### ABOUT THE MODEL

James Cameron is known for creating fantastic vehicles for his movies, and **Avatar** was no exception. One of the most inventive vehicles in the film was the Samson Battle Helicopter. This model of it features two double counter-rotating rotors inside pods that can be tilted by a two-way mechanical joystick, cockpit doors that open, and spring-loaded cargo doors. The finished model is truly massive, and its design was partially computer-aided.

### CHALLENGES

The rotor pods were the primary challenge because of the complexity involved in driving two counter-rotating rotors. In the end, the rotor pads proved so heavy that side supports were needed to keep them properly aligned.

### THE ORIGINAL

The SA-2 Samson Battle Helicopter is a twin-turbine, ducted-rotor helicopter that functions as a futuristic counterpart of the Black Hawk helicopter by carrying personnel and dropping supplies under combat conditions.

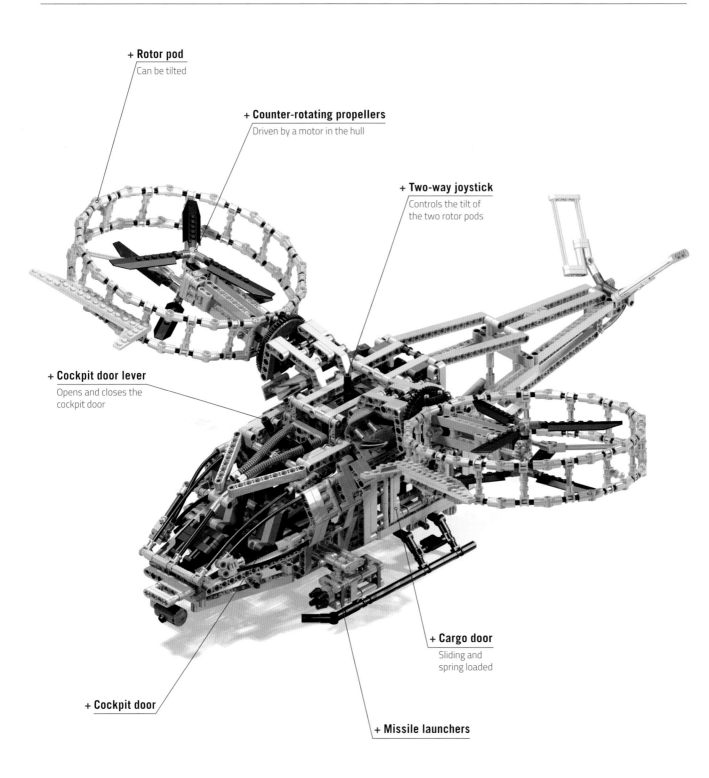

+ **Rotor pod**
Can be tilted

+ **Counter-rotating propellers**
Driven by a motor in the hull

+ **Two-way joystick**
Controls the tilt of
the two rotor pods

+ **Cockpit door lever**
Opens and closes the
cockpit door

+ **Cargo door**
Sliding and
spring loaded

+ **Cockpit door**

+ **Missile launchers**

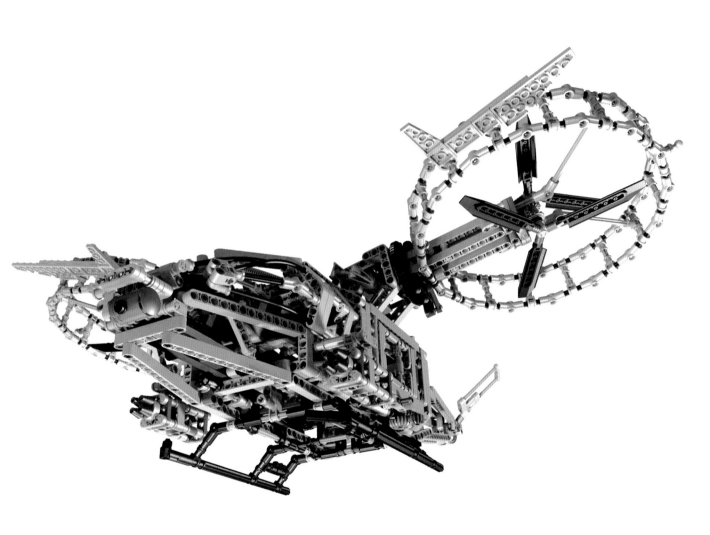

When this model is displayed in public, children often ask whether it can really fly. The builder's answer is yes, but it depends on how far you throw it.

# SPITFIRE

*Thirdwigg (2012)*

## SPECIFICATIONS

| | |
|---|---|
| LENGTH | **35.4"** |
| WIDTH | **30.3"** |
| HEIGHT | **8.7"** |
| PIECES | **3,074** |

## ABOUT THE MODEL

This 1:12 scale model of the iconic British Spitfire plane is a splendid combination of looks and function. The World War II—era plane has a motorized propeller, an adjustable pitch, a retractable landing gear, and even a V12 piston engine inside the nose. The cockpit features a working joystick that operates the elevator and ailerons, two pedals that move the rudder, and a lever that deploys the flaps. All these functions are integrated into a sturdy skeleton of Technic bricks covered with plates that mimic the camouflage of the original plane. Even the markings are made with LEGO pieces and recreate a specific machine: a 1941 Spitfire Mk IIa from the 312th Squadron, which was flown by Czech pilot Lt. Tomas Vybiral.

## CHALLENGES

The main challenge is that planes are made of many long and thin parts. That means the model needs a sturdy skeleton; for all this model's size and weight there is actually little space for the functions inside. This model, which took seven months to develop, uses a battery box as part of the structure, and its massive weight of 14.3 lb (6.5 kg) rests on three small retractable wheels. The body skin was a separate challenge, as it required many rare LEGO pieces in unusual colors and needed to be removable to allow access to the model's innards.

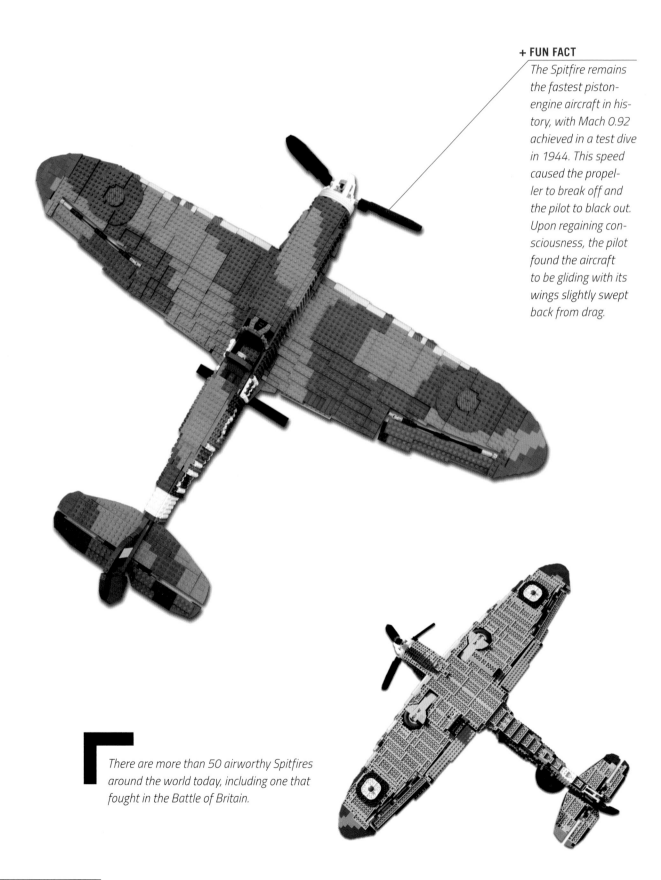

The Spitfire remains the fastest piston-engine aircraft in history, with Mach 0.92 achieved in a test dive in 1944. This speed caused the propeller to break off and the pilot to black out. Upon regaining consciousness, the pilot found the aircraft to be gliding with its wings slightly swept back from drag.

There are more than 50 airworthy Spitfires around the world today, including one that fought in the Battle of Britain.

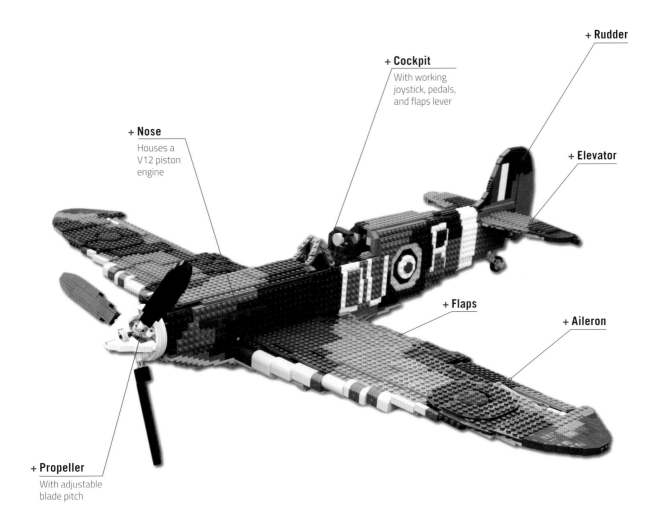

**+ Rudder**

**+ Cockpit**
With working
joystick, pedals,
and flaps lever

**+ Nose**
Houses a
V12 piston
engine

**+ Elevator**

**+ Flaps**

**+ Aileron**

**+ Propeller**
With adjustable
blade pitch

## THE ORIGINAL

The Spitfire was the most-produced British
fighter aircraft of World War II. Serving as a
short-range interceptor, it proved essential
during the Battle of Britain because it out-
performed its German counterparts. It was
produced and continually improved throughout
the war and saw action in Europe, the Pacific,
and Asia. While somewhat limited in its role as
a point-defense aircraft, the Spitfire is considered one of World War II's best fighters, along
with the P-51 Mustang and the Focke-Wulf
Fw 190.

# T-47 AIRSPEEDER "REBEL SNOWSPEEDER"

*drakmin (2014)*

## ABOUT THE MODEL

This model of the Rebel Snowspeeder, featured in **Star Wars: The Empire Strikes Back**, is an example of the difficult art of building an authentic-looking vehicle with studless LEGO pieces. Unlike official Technic models, it comes with a body that is smooth and free of holes but at the same time filled with functions. Incredibly, the functional elements—including working airbrakes, rear flaps, and a rear harpoon that can be aimed—are controlled by two joysticks inside the cockpit. The complexity of this model may be the reason why it took three years to develop.

## THE ORIGINAL

The T-47 Airspeeder, originally a cargo-handling transport, was modified into a combat-ready craft that became commonly known as the Snowspeeder, following its use in the battle of Hoth. Capable of high-speed maneuvers but lacking shields, the Snowspeeder relies on its agility to dodge incoming fire. Its cargo harpoon is a formidable weapon, famously used to tie the legs of advancing AT-ATs.

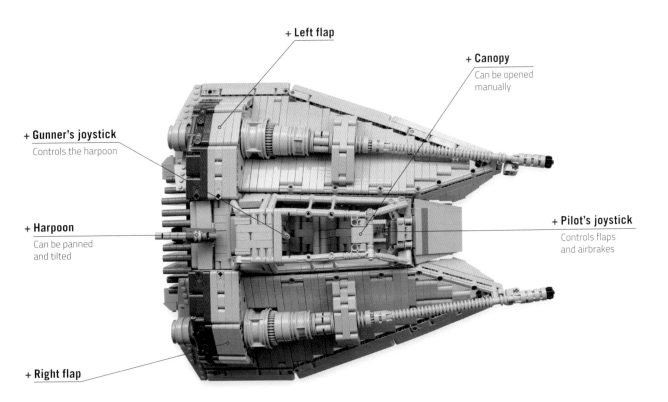

+ **Left flap**

+ **Canopy**
Can be opened
manually

+ **Gunner's joystick**
Controls the harpoon

+ **Harpoon**
Can be panned
and tilted

+ **Pilot's joystick**
Controls flaps
and airbrakes

+ **Right flap**

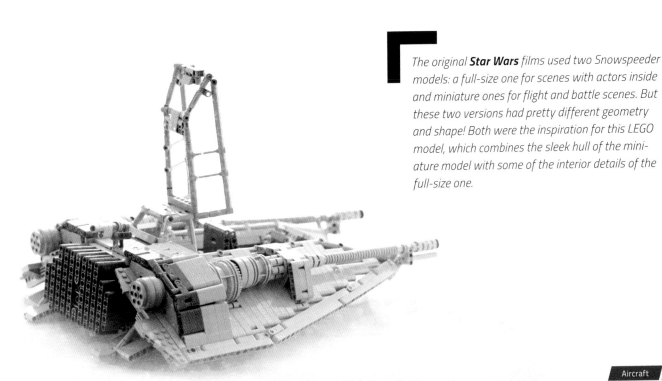

The original **Star Wars** films used two Snowspeeder
models: a full-size one for scenes with actors inside
and miniature ones for flight and battle scenes. But
these two versions had pretty different geometry
and shape! Both were the inspiration for this LEGO
model, which combines the sleek hull of the miniature model with some of the interior details of the
full-size one.

# AUTOMOTIVE

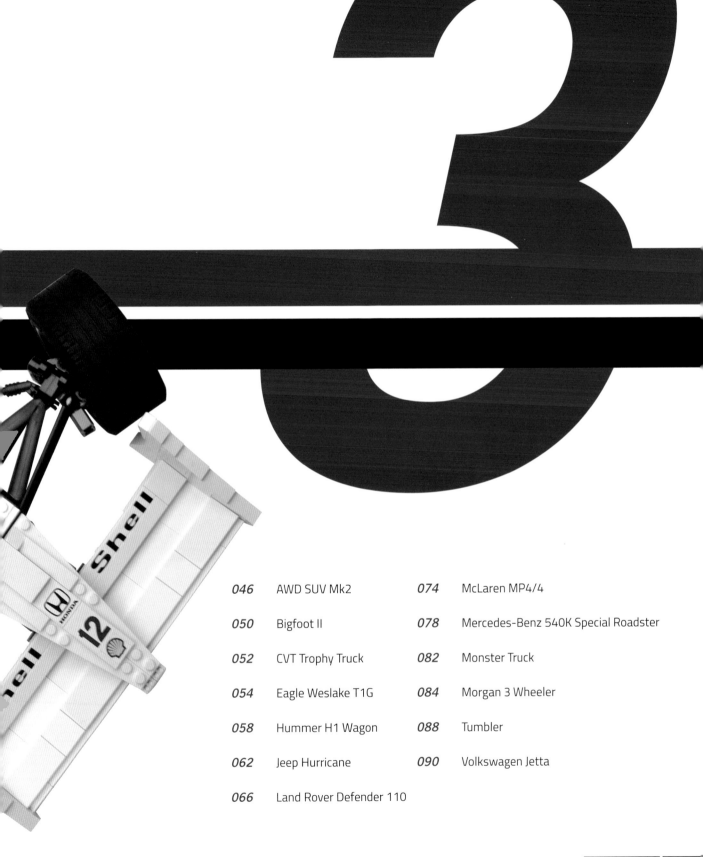

# 3

# AWD SUV MK2

*Madoca (2013)*

## ABOUT THE MODEL

This generic SUV model evokes the look of a modern Land Rover and includes some fantastic functions. The model is all-wheel drive and features a full suspension that can be raised and lowered, a two-speed transmission, and working headlights and taillights. In addition, the doors, trunk, and hood can all be opened. With plenty of torque from two PF XL motors and a ground clearance that can be increased as needed, the model is more than able to run a serious obstacle course.

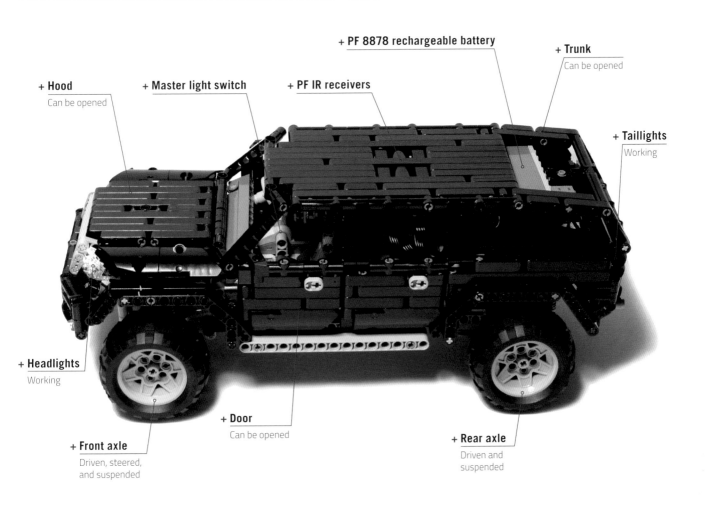

+ **Hood**
Can be opened

+ **Master light switch**

+ **PF IR receivers**

+ **PF 8878 rechargeable battery**

+ **Trunk**
Can be opened

+ **Taillights**
Working

+ **Headlights**
Working

+ **Door**
Can be opened

+ **Front axle**
Driven, steered,
and suspended

+ **Rear axle**
Driven and
suspended

## SPECIFICATIONS

| | |
|---|---|
| LENGTH | **14.4"** |
| WIDTH | **6.7"** |
| HEIGHT | **5.3"** |
| PIECES | **1,375** |

## CHALLENGES

The primary challenge was making both the rear and front suspensions adjustable within the confined space imposed by the model's relatively small size. To achieve this, the model has two small linear actuators that lower or raise the upper mounting points of the suspension's shock absorbers.

# BIGFOOT II

*Andrea Grazi (2003)*

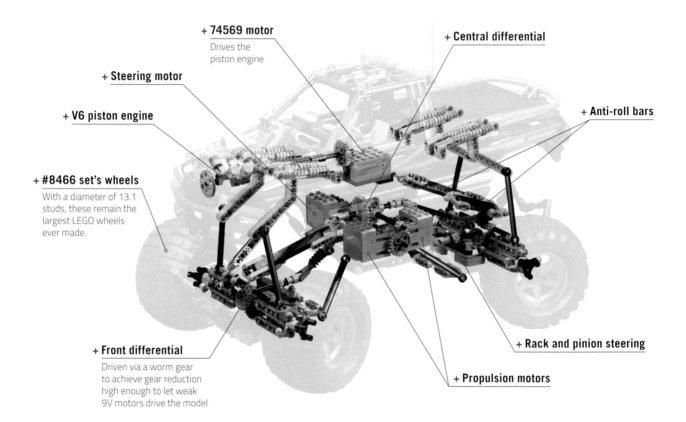

+ **74569 motor**
Drives the
piston engine

+ **Steering motor**

+ **V6 piston engine**

+ **#8466 set's wheels**
With a diameter of 13.1
studs, these remain the
largest LEGO wheels
ever made.

+ **Central differential**

+ **Anti-roll bars**

+ **Front differential**
Driven via a worm gear
to achieve gear reduction
high enough to let weak
9V motors drive the model

+ **Rack and pinion steering**

+ **Propulsion motors**

## ABOUT THE MODEL

This model demonstrates the pinnacle of LEGO Technic engineering in the era before Power Functions. It features a long-travel suspension with two floating axles, all-wheel drive with a central differential, and all-wheel steering. A separate motor drives the V6 piston engine, and the model is relatively lightweight because it doesn't have any onboard power supply. In the early 2000s, LEGO motors were controlled via wires using handheld battery boxes, and it took one battery box for each motor! In many ways, this model can be seen as a predecessor of the famous LEGO Technic Crawler set (#9398).

## CHALLENGES

There were a number of challenges in 2003 that don't affect modern LEGO builders: creating a complex and robust suspension system with a limited palette of pieces, using differentials that were difficult to connect to a drivetrain, and propelling the model with motors substantially weaker than the ones available today.

## SPECIFICATIONS

| | |
|---|---|
| LENGTH | **16.2"** |
| WIDTH | **10.9"** |
| HEIGHT | **11.3"** |
| PIECES | **1,816** |

# CVT TROPHY TRUCK

*Nico71 (2013)*

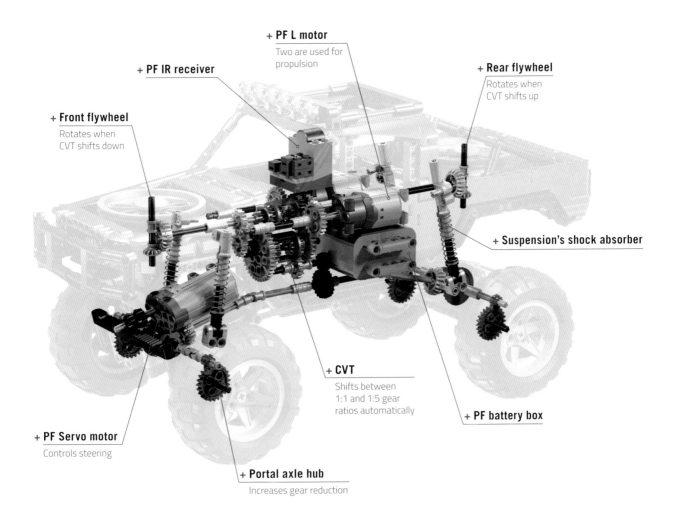

**+ PF L motor**
Two are used for propulsion

**+ PF IR receiver**

**+ Rear flywheel**
Rotates when CVT shifts up

**+ Front flywheel**
Rotates when CVT shifts down

**+ Suspension's shock absorber**

**+ CVT**
Shifts between 1:1 and 1:5 gear ratios automatically

**+ PF battery box**

**+ PF Servo motor**
Controls steering

**+ Portal axle hub**
Increases gear reduction

## SPECIFICATIONS

| | |
|---|---|
| LENGTH | **16.2"** |
| WIDTH | **8.3"** |
| HEIGHT | **8.8"** |
| PIECES | **1,121** |

## ABOUT THE MODEL

This off-road truck model was built to demonstrate a *continuously variable transmission (CVT),* which is a stepless transmission system that automatically adjusts its gear ratio between two extremes. In this case, the ratio can be changed from 1:1 to 1:5, depending on how much resistance the wheels are generating. This means the CVT can effectively increase torque five times if needed, with the two PF L propulsion motors running at constant speed at all times. Other functions include steering with Ackermann geometry, a return-to-center function, and a long-travel suspension with live axles.

## CHALLENGES

CVTs built with LEGO pieces tend to shift in a rapid, jerky way. To counter this tendency, two horizontal flywheels add latency and make the CVT work more smoothly. One is connected to the CVT's 1:1 output, and the other is connected to the 1:5 output. By observing the flywheels as they rotate, it is possible to determine the current gear ratio.

# EAGLE WESLAKE T1G

*RoscoPC (2013)*

## SPECIFICATIONS

| | |
|---|---|
| LENGTH | **21.1"** |
| WIDTH | **10.1"** |
| HEIGHT | **4"** |
| PIECES | **1,495** |

## ABOUT THE MODEL

This model has all the functions typical of a large nonmotorized Formula 1 car model, with its working steering, drivetrain with a V12 piston engine, and suspension system. But this Eagle Weslake was built with a focus on aesthetics. The massive engine helps make the chassis rigid, and the construction of the suspension is accurate, including stabilizer bars and bent rocking arms. The smooth body is carefully shaped with a variety of small elements, which is why, for all its simple looks, the model uses almost 1,500 LEGO pieces.

## CHALLENGES

The primary challenge followed the choice of the dark blue livery; it was difficult to model the body, and particularly the nose, using the limited array of LEGO pieces available in this color. Another challenge was to make the chassis rigid, but in the end, the model is sturdy enough to be wielded like a club.

## THE ORIGINAL

The Eagle Weslake T1G was a Formula 1 car introduced at the 1966 Belgian Grand Prix. With its sleek blue body, white nose, and lone white stripe, it is widely considered to be one of the most beautiful Grand Prix cars ever made. Initially an aluminum-based construction, it quickly saw the introduction of advanced metal alloys, including multiple magnesium sheets in the panelwork. Sophisticated and powerful, the car was hindered only by the unreliability of its engine, which allowed it to win the Belgian Grand Prix only once, in 1967. That makes it the only American car to win a Formula 1 Grand Prix ever.

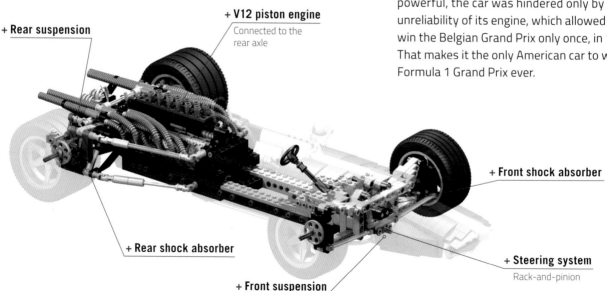

+ **Rear suspension**

+ **V12 piston engine**
Connected to the rear axle

+ **Front shock absorber**

+ **Rear shock absorber**

+ **Front suspension**

+ **Steering system**
Rack-and-pinion

The magnesium extensively used in the car's panel-work was highly flammable, which was obvious to the driver Dan Gurney. On several occasions he drove without safety belts because they would slow down his escape in case of fire.

+ FUN FACT

*Some dark blue pieces proved so rare and expensive that changes to the design were necessary. The dark blue 1×2 inverted slope, which appears only once in a single LEGO set, sells for as much as $25 a piece.*

# HUMMER H1 WAGON

*Sariel (2014)*

**SPECIFICATIONS**

LENGTH **21.4"**

WIDTH **9.5"**

HEIGHT **9.5"**

PIECES **~3,000**

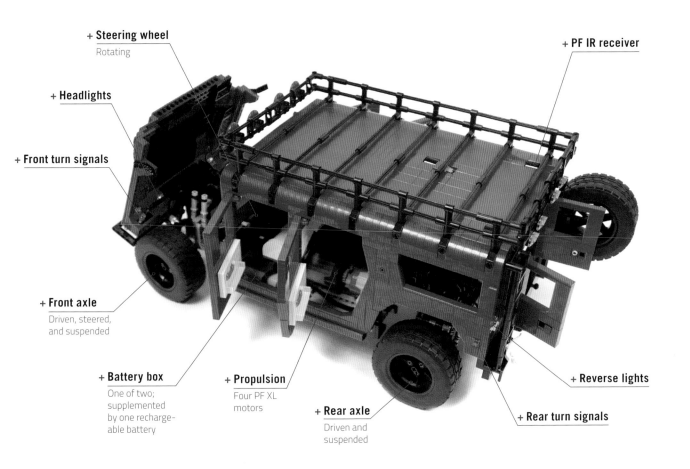

+ **Steering wheel**
Rotating

+ **Headlights**

+ **Front turn signals**

+ **Front axle**
Driven, steered,
and suspended

+ **Battery box**
One of two;
supplemented
by one recharge-
able battery

+ **Propulsion**
Four PF XL
motors

+ **Rear axle**
Driven and
suspended

+ **PF IR receiver**

+ **Reverse lights**

+ **Rear turn signals**

### ABOUT THE MODEL

My third take on the iconic Hummer, this massive model features a 4×4 drive with portal
axles, a full independent suspension, two-
speed remote-controlled transmission, and
a rotating steering wheel, as well as working
headlights, taillights, automated turn signals,
and reverse lights. To top it all off, it even has
working windshield wipers.

The doors, trunk, and hood can be opened,
and the 9 lb (4 kg) model is propelled by four PF
XL motors and powered by two battery boxes
and one rechargeable battery. All power supplies are located under the front cabin for better
weight distribution.

### CHALLENGES

The model's weight presented a serious challenge. It was difficult to build a robust independent suspension that would keep such a heavy
model in balance, especially with the wagon
body being very back-heavy. In the end, the
front and real suspensions vary a lot, and
the chassis is greatly reinforced to prevent it
from bending despite the model's enormous
wheelbase.

*Serving for more than 30 years, the original Humvee has become perhaps the world's most popular military vehicle, with an estimated 160,000 units operated by 36 countries. Ten thousand Humvees were used in the Iraq War alone. An extremely versatile vehicle, the Humvee is produced in more than 40 variants for the US Army alone, not including international variants. There are classified variants too, operated by special forces, like the Polish GROM unit.*

## + FUN FACT

The model has enough ground clearance to drive over a LEGO mini-figure standing upright.

### THE ORIGINAL

The Hummer H1 is a civilian variant of the Humvee, a high-mobility vehicle used by militaries since 1984. Prominently used in Operation Desert Storm and popularized by Arnold Schwarzenegger, the vehicle was eventually released to the civilian market by popular demand. Coming off the same assembly line as the Humvee and fitted with the same military-grade features, the Hummer is able to outperform most regular cars, while its sheer size and crude silhouette make it stand out. Hummers are loved or hated but rarely ignored.

# JEEP HURRICANE

*NKubate (2011)*

## ABOUT THE MODEL

This model attempts to re-create the advanced functions of a Jeep concept vehicle. It includes a fully independent suspension, all-wheel drive, two V8 piston engines, a drivetrain that allows normal drive or skid steering, and two steering modes (regular and "toe-in" steering). The completed model includes approximately 60 gear wheels and is built so densely that it weighs 4.4 lb (2 kg), despite being rather compact.

*To reduce fuel consumption, the original Jeep Hurricane features an automatic engine displacement system that can disengage four-piston engine blocks in either of its two HEMIs depending on how much power is needed. This means that the Hurricane doesn't need all its 16 pistons to work; it can run on a single engine when on a dirt road and on just 4 pistons when on a highway.*

## SPECIFICATIONS

| | |
|---|---|
| LENGTH | **17.2"** |
| WIDTH | **10"** |
| HEIGHT | **7.4"** |
| PIECES | **1,724** |

### THE ORIGINAL

Jeep Hurricane is a concept vehicle designed as the ultimate off-road machine, but it was never mass produced. Its unusual two V8 HEMI engine design is just the beginning; the drivetrain and steering system make this car unique. Each wheel steers independently, enabling regular steering, crab steering, and even "toe-in" steering that makes the Hurricane turn on the spot. It can also skid steer, thanks to separate driveshafts for the right and left wheels.

### CHALLENGES

The steering system presented the primary challenge. Complex mechanical solutions were needed to keep all wheels aligned at all times, regardless of the steering mode being used. A drivetrain that would allow skid steering was challenging too, requiring the differentials to be installed not between the wheels on a single axle but between the two left and two right wheels.

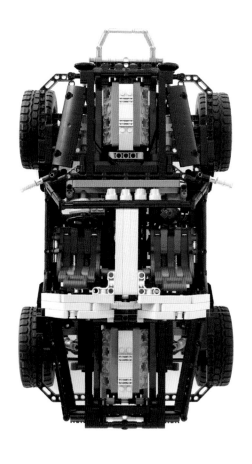

# LAND ROVER DEFENDER 110

*Sheepo (2012)*

SPECIFICATIONS

| | |
|---|---|
| LENGTH | **20.4"** |
| WIDTH | **9.4"** |
| HEIGHT | **9.6"** |
| PIECES | **3,437** |

## ABOUT THE MODEL

This model of the Land Rover Defender is built around a double beam frame that is fitted with two live axles, an FWD/AWD selector, four disc brakes, a straight-four piston engine, a 5+R sequential remote-controlled transmission with an automatic clutch, and a gear reducer that can be engaged when extra torque is needed, just like in the real car. The body is entirely built using modern beams and panels, with doors, a hood, and a tailgate that open, and it can be removed from the chassis in one piece.

## CHALLENGES

While the bodywork was quite easy, the chassis posed a serious challenge. It was difficult to make such complex elements as the transmission, the clutch, the reducer, and the drive selector fit between the frame beams. The final chassis design is modular; the transmission and both axles can be easily detached.

### THE ORIGINAL

The Defender is a base vehicle line of the Land Rover company, first introduced in 1948 and with subsequent models remaining in production until now. Designed as a versatile AWD utility vehicle, it is known for reliability and is operated by civilians, police forces, and military around the world. Proven in many of the world's most challenging environments, from the jungle to arctic regions and even mountain summits (there's one Land Rover left atop Mount Elbrus), the Defender enjoys the reputation of a car you can drive anywhere and mend with a hammer and spanner.

*It's estimated that more than 60 percent of all Land Rovers ever manufactured are still in working condition.*

*This impressive model, which took five months to develop, was submitted to the LEGO CUUSOO program (LEGO's fan-submitted design program, now called LEGO Ideas), where it enjoyed massive support and attention of the world-wide Technic community. After gathering the 10,000 votes required, it was rejected by the LEGO review team for undisclosed reasons.*

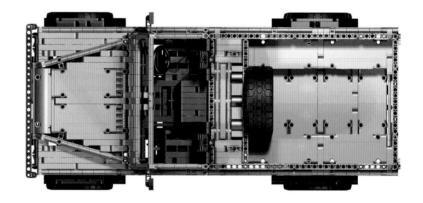

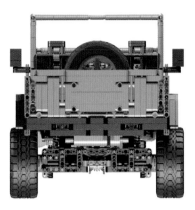

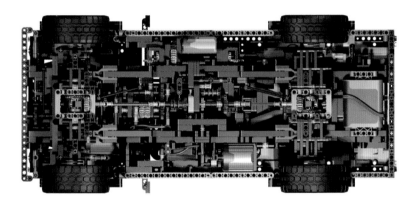

# MCLAREN MP4/4

*RoscoPC (2011)*

## SPECIFICATIONS

| | |
|---|---|
| LENGTH | **21.7"** |
| WIDTH | **10.1"** |
| HEIGHT | **4.9"** |
| PIECES | **1,584** |

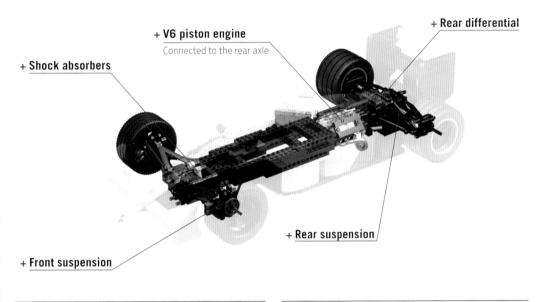

+ **V6 piston engine**
Connected to the rear axle

+ **Rear differential**

+ **Shock absorbers**

+ **Rear suspension**

+ **Front suspension**

## ABOUT THE MODEL

This model focuses on re-creating the distinctive original livery of one of the world's most famous Formula 1 cars. The complex red-and-white scheme uses only LEGO pieces, without resorting to stickers. The model's functions include a working suspension in pushrod configuration, a steering system, and a working drivetrain with the V6 piston engine. The car's silhouette is not only exceptionally low but is also full of slim and tapering shapes that were difficult to model without compromising the livery. The model's slim profile also severely limited the amount of space inside the chassis. In the end, there are almost 1,600 LEGO pieces inside this relatively small and compact model.

## CHALLENGES

The main challenge was making the front and rear suspension compact enough to fit inside the slim nose and the low rear end of the car, with the latter partially occupied by a rising lower diffuser.

## THE ORIGINAL

The McLaren MP4/4 was a Formula 1 car introduced in the 1988 Formula 1 season, during which it won all but one race, making it one of the most successful Formula 1 cars in history. Its winning streak was interrupted only when Formula 1 regulations banned turbo engines outright, starting with the 1989 season.

The car's low silhouette and the FIA safety rules (which require the driver's helmet to remain at a certain distance below the roll bar) forced the drivers into a liedown position rather than the then-conventional upright seating position of Grand Prix cars.

# MERCEDES-BENZ 540K SPECIAL ROADSTER

*Sariel (2012)*

## SPECIFICATIONS

| | |
|---|---|
| LENGTH | **17"** |
| WIDTH | **5.7"** |
| HEIGHT | **5"** |
| PIECES | **1,018** |

### ABOUT THE MODEL

This is a relatively simple model, combining a brick-built chassis with two Power Functions motors—a PF L motor for driving and a Servo motor for steering. The detailed body consists of more than 1,000 LEGO pieces, but there's no suspension system, and the technical sophistication ends with a working steering wheel. The resulting model includes just 8 gear wheels but has more than 80 curved slopes as well as a few custom-chromed pieces.

Because the original car was often customized to meet customers' needs, the model combines the details of several Special Roadsters rather than trying to re-create a single car.

At 2.8 lb (1.27 kg), this model's powerful, large motor, the dead-simple drivetrain, and the quickly working servo motor make it agile and fun to drive. The cabin is just the right size for a hamster to take a ride.

### THE ORIGINAL

Introduced in 1936, the 540K represents the pinnacle of German engineering and craftsmanship at the time. Designed as a top-of-the-line luxury car, it was built only by special order. There were 10 body variants to choose from, and the Special Roadster, styled by Hermann Ahrens, was one of them. Only 25 Special Roadsters were built, and only 10 of them exist today.

The car is often called the most beautiful German supercar of all time, and its performance is every bit as good as its looks. It is fitted with a four-wheel independent suspension and a four-speed gearbox with the three top gears synchronized. Underneath the long, sleek hood, there is an inline-8 engine capable of producing 115 hp, and a supercharger can be engaged for short periods of time to increase the power to 180 hp.

The timeless beauty and technical sophistication well ahead of its time make the 540K Special Roadster one of the most sought-after supercars in the world.

**+ Hood**

Curved and tapering.
Made of many curved
slopes offset by half
a stud.

**+ Door**

Can be opened
manually

**+ IR receiver**

The only one in
the entire model

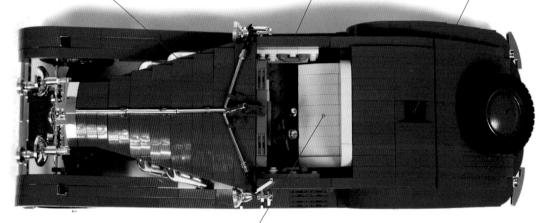

**+ Cabin**

With detailed interior
and enough space for
two hamsters. If only
they agreed to wear
seat belts!

**+ PF Servo motor**

Controls steering

**+ PF L motor**

Serves as propulsion,
geared at 1:1

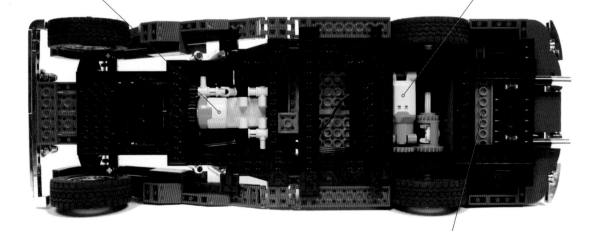

**+ PF battery**

The 8878 rechargeable
battery is the model's
sole power source.

+ FUN FACT

The round LEGO piece used to imitate the Mercedes emblem on the hood is actually the One Ring from the Lord of the Rings LEGO sets. Sauron just called; he wants it back.

Jack Warner, head of Warner Bros., kept a 540K Special Roadster for more than 10 years. After being restored to perfect condition, it was recently sold for $3.63 million.

# MONSTER TRUCK

*Crowkillers (2013)*

### ABOUT THE MODEL

This small-scale model of a monster truck combines impressive functionality with a sleek body that can be removed. The chassis features an independent suspension with floating axles, a 4×4 drive with a central differential, all-wheel steering, and a V8 piston engine, which is massive at this scale. The body is skillfully shaped with just a handful of small panels and axles, the doors can be opened, and the rims are custom chromed.

### CHALLENGES

It was difficult to fit such a complex drivetrain and suspension system into a chassis that is just 13 studs wide. The finished model relies on LEGO CV joints and short shock absorbers to make it possible.

## SPECIFICATIONS

| | |
|---|---|
| LENGTH | **9.8"** |
| WIDTH | **7.2"** |
| HEIGHT | **6.9"** |
| PIECES | **~700** |

# MORGAN 3 WHEELER

*Nico71 (2012)*

### ABOUT THE MODEL

Despite being styled as a classic 1930s car, the Morgan 3 Wheeler debuted in 2011. This model of one of Britain's more exotic cars features a four-speed transmission, a working suspension, a steering and piston engine with pushrods and rockers, an opening trunk, and a removable hood. Because it's not motorized, the Morgan is able to pack all these functions into a compact chassis. The transmission produces gear ratios from 1:1 to 1:6, the suspension consists of double wishbones in the front and a swinging arm in the back, and the fully exposed engine is spectacular to watch thanks to its many moving parts.

### CHALLENGES

Strangely enough, the main challenge with this model was its body. The original Morgan car has few straight lines, with a rounded, cigar-shaped silhouette. The thin, tapering back end proved to be most difficult, especially with the opening trunk.

### THE ORIGINAL

Founded in 1910, the British Morgan Motor Company specializes in building sport cars. Well known in Europe and the United States (at one point the US market absorbed 85 percent of the company's production), all Morgan cars are assembled by hand. The modern but vintage-looking results are so popular with classic-car enthusiasts that the waiting list is roughly two years long.

## SPECIFICATIONS

| | |
|---|---:|
| LENGTH | **16.4"** |
| WIDTH | **8.2"** |
| HEIGHT | **5"** |
| PIECES | **970** |

### + FUN FACT

*Morgans have a decades-long history of being non-compliant with US vehicle safety regulations, making them difficult to import. A variety of workarounds counter this, including regulating the Morgan 3 Wheeler as a motorcycle.*

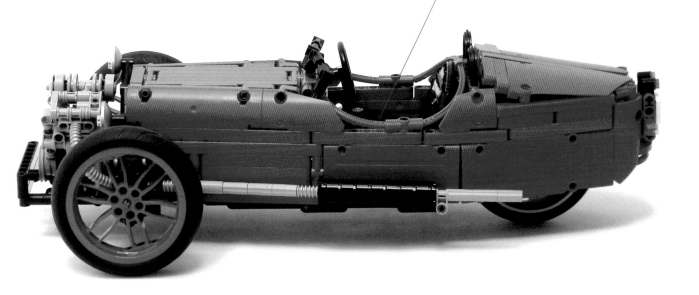

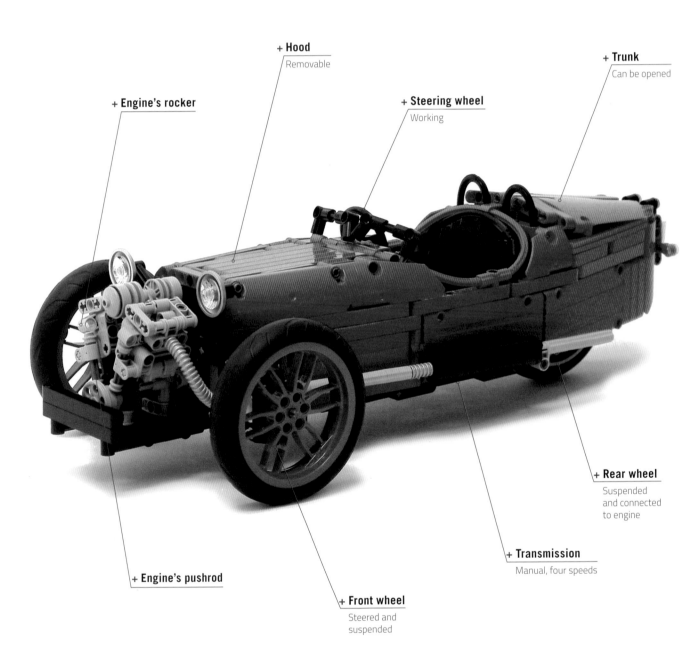

+ **Hood**
Removable

+ **Trunk**
Can be opened

+ **Engine's rocker**

+ **Steering wheel**
Working

+ **Rear wheel**
Suspended
and connected
to engine

+ **Transmission**
Manual, four speeds

+ **Engine's pushrod**

+ **Front wheel**
Steered and
suspended

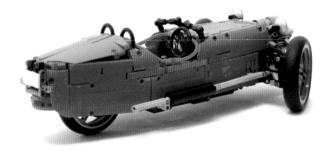

The Morgan 3 Wheeler is a re-imagining of the three-wheelers Morgan produced from 1910 to 1953. It offers performance on par with a regular sport car because of its minimal weight.

# TUMBLER

*Sariel (2012)*

## SPECIFICATIONS

| | |
|---|---|
| LENGTH | **17.6"** |
| WIDTH | **10.7"** |
| HEIGHT | **6"** |
| PIECES | **~2,000** |

*The street-ready Tumbler developed for **Batman Begins** proved so fast and agile that the Mercedes-Benz ML55 AMG cars used as camera carriers for chase sequences struggled to keep up with it. It was so problematic that the two later Batman movies made use of supercharged ML55s instead.*

## ABOUT THE MODEL

At 17.6 inches (45 cm) long and 5.16 lb (2.34 kg), this Tumbler is full of details and functions. It also proved to be a challenging build, years in the making. Its propulsion system, with two LEGO RC motors coupled with an adder, proved to be the easiest bit, compared with the complex suspension and steering system.

The finished model is fast, agile, and capable of making sharp turns with a visible body roll, just like the "real thing" seen in the movies. The model includes an array of LEGO LEDs and a rotating backlit exhaust flame in the jet nozzle. There are also two spring cannons in the front that can be triggered remotely. All of this functionality is powered by just an RC unit and a separate battery box for the LEDs.

The model was called "the best R/C Batmobile ever" by *Top Gear.*

## CHALLENGES

For all of the model's complexity, the front suspension proved to be the biggest challenge. It had to be independent, steerable, compact, and robust enough to support a heavy, fast model. The final version is supported by a rigid structure made of upside-down Technic bricks, forming thin suspension arms connected to the central body frame at five points each. It's robust enough to let the model perform just like the original Tumbler—only at smaller scale.

It wasn't until two years later that LEGO released set #76023, which included a faithful reproduction of the original Tumbler's front tires at a scale matching this model. But at the time this model was created, the front wheels had to be improvised.

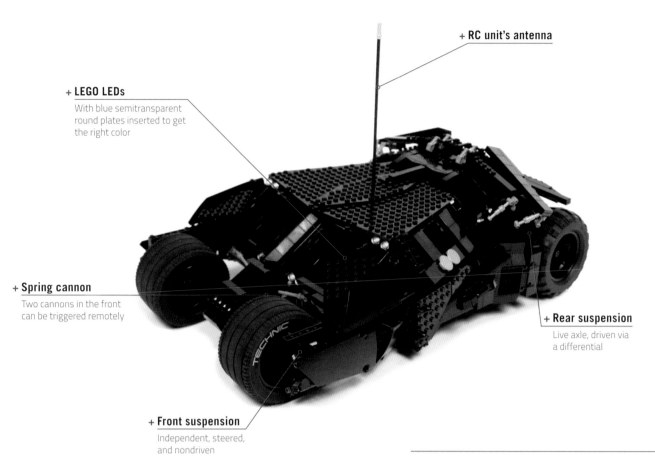

**+ RC unit's antenna**

**+ LEGO LEDs**
With blue semitransparent
round plates inserted to get
the right color

**+ Spring cannon**
Two cannons in the front
can be triggered remotely

**+ Rear suspension**
Live axle, driven via
a differential

**+ Front suspension**
Independent, steered,
and nondriven

**+ FUN FACT**
*The spinning orange
flame piece in the jet
nozzle is a tribute to
the LEGO Batmobile
set (#7784) released
in 2006.*

## THE ORIGINAL

The tank-like Tumbler is a new kind of
Batmobile created for Christopher Nolan's
movie trilogy. It leaves behind the sleek, futuristic vehicles from the earlier Batman movies
in order to fit with the realistic, gritty tone.
And perhaps in keeping with the new, realistic
approach, the vehicle is not even once called a
***Batmobile***.

The production crew spent nine months and
several million dollars developing a real, street-
ready Tumbler from scratch. The resulting vehicle was a 2.5-ton monster with a 500 hp engine
and a truck axle in the rear, using suspension
from Baja trucks to make 39 foot (12 meter)
jumps possible. Three more were built, cost-
ing $250,000 each, including one for close-up
shots and another with a propane-powered
mock jet engine. A 1:6 scale radio-controlled
model was also built and used for roof-jumping
sequences. Finally, it took six months of practice for each of the stunt drivers to really master driving the Tumbler.

# VOLKSWAGEN JETTA

*Spiderbrick (2012)*

### ABOUT THE MODEL

While most supercar-scale models re-create exotic fuel eaters, this one stands out by focusing on an everyday sedan. With a fully independent suspension, front-wheel drive, a manual 5+R transmission, a piston engine, and a remote control, the Volkswagen Jetta is every bit as advanced as its flashier rivals, despite being a daily driver.

### CHALLENGES

The model's front end was quite challenging. It's far from spacious, but it includes many functions. The front-wheel drive, transverse piston engine, transmission, and motorized steering with a working steering wheel all fit and work together well.

### SPECIFICATIONS

| | |
|---|---|
| LENGTH | **18.1"** |
| WIDTH | **7.9"** |
| HEIGHT | **4.7"** |
| PIECES | **1,650** |

## THE ORIGINAL

Created in 1979 by simply adding a trunk to the hatchback Volkswagen Golf, the Jetta has grown considerably over its six generations. Its latest incarnation is a luxurious small saloon car, available with engines up to 170 hp, a turbocharger, and power steering in some models. Its high-performance version, the Jetta R, is intended as a rival for sportier sedans such as the Mitsubishi Lancer Evolution.

# CONSTRUCTION EQUIPMENT

# ARTICULATED HAULER 6×6

*Designer-Han (2011)*

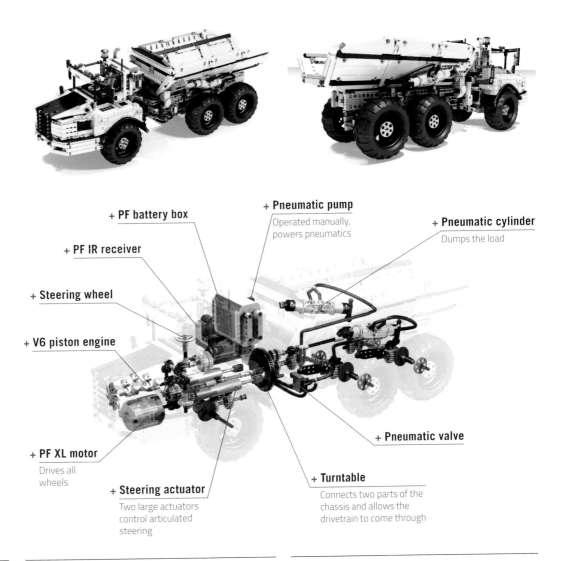

+ **PF battery box**

+ **PF IR receiver**

+ **Steering wheel**

+ **V6 piston engine**

+ **Pneumatic pump**
Operated manually,
powers pneumatics

+ **Pneumatic cylinder**
Dumps the load

+ **PF XL motor**
Drives all
wheels

+ **Steering actuator**
Two large actuators
control articulated
steering

+ **Turntable**
Connects two parts of the
chassis and allows the
drivetrain to come through

+ **Pneumatic valve**

## SPECIFICATIONS

| | |
|---|---|
| LENGTH | **22.7"** |
| WIDTH | **6.9"** |
| HEIGHT | **8.3"** |
| PIECES | **2,140** |

## ABOUT THE MODEL

Inspired by the Volvo A40D 6×6 hauler, this model features an impressive drivetrain and suspension system, as well as articulated steering, a working steering wheel, a pneumatic dumping mechanism, a V6 piston engine that can be accessed by opening the hood and the front grille, and working headlights. The engine even has a cooling fan!

## CHALLENGES

The original Volvo hauler comes equipped with a unique three-point suspension system that allows independent movement of all four rear wheels. It was necessary to mount axles on ball joints and to create extending sections in the driveshaft to re-create this suspension in the model.

## THE ORIGINAL

The A40D hauler manufactured by the Swedish Volvo company is driven by a 420 hp V6 engine and steered with hydraulics. It's specifically designed for rough terrain rather than speed, with a top speed of 35 mph (55 km/h). It can haul 37 tons that can be dumped completely in just 12 seconds.

# CATERPILLAR 7495 HF

*Konajra (2014)*

This power shovel design originates from the Bucyrus company, which is now owned by Caterpillar. The Caterpillar 7495 HF is effectively an original Bucyrus 495HF shovel in Caterpillar livery.

## ABOUT THE MODEL

This minifig-scale model of the Caterpillar 7495 HF power shovel blends functionality with carefully detailed styling. It's massive, with a superstructure sitting on a roller bearing made with round bricks, as well as a chassis propelled by four PF XL motors. Four PF M motors operate the massive bucket, while two PF XL motors are used for slewing. The bucket was custom-built rather than using a LEGO ready-made one; it was carefully shaped and equipped with a bottom that opens via remote control. There are even custom segments attached on top of the large Technic tracks for a more accurate look.

## CHALLENGES

Other than the challenges resulting from the sheer size and weight of the model, such as smooth driving and slewing, the main challenge was the bucket. It had to be lifted and tilted, with its bottom safely closed shut, but able to be opened at any moment—all with the remote control.

## THE ORIGINAL

The Caterpillar 7495 HF is an electric rope shovel, also called a **power shovel**. Massive machines such as this are used for high-capacity digging and loading operations in strip mines. With a bucket capacity upward of 110 tons and weighing nearly 1,500 tons itself, the monstrous 7495HF can effectively replace a whole fleet of regular-sized excavators.

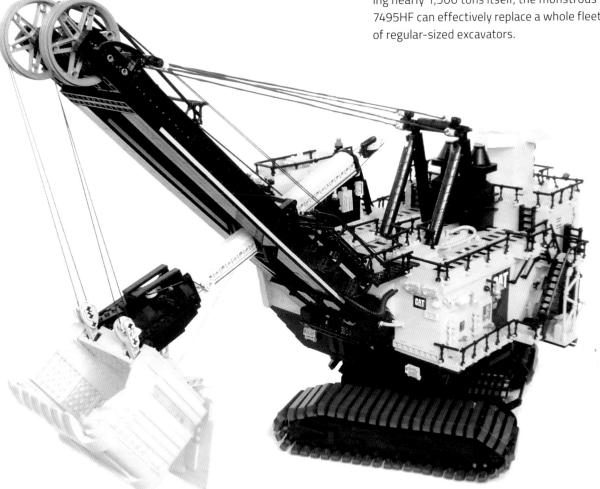

# CATERPILLAR D9T

ZED (2013)

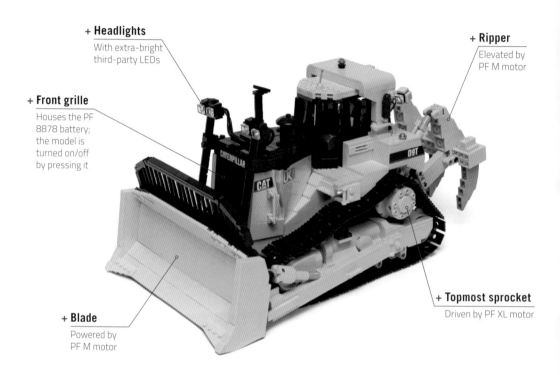

**+ Headlights**
With extra-bright third-party LEDs

**+ Ripper**
Elevated by PF M motor

**+ Front grille**
Houses the PF 8878 battery; the model is turned on/off by pressing it

**+ Blade**
Powered by PF M motor

**+ Topmost sprocket**
Driven by PF XL motor

## ABOUT THE MODEL

This 1:22 model of the Caterpillar D9T tracked dozer looks almost too good to be Technic. It comes with motorized tracks, an elevated dozer blade and ripper, and an array of third-party LEDs. Two PF XL motors give it plenty of power to push stuff around, and the PF M motor controlling the ripper has enough force to lift the model's back end off the ground. Because the looks of the model were extremely important, the blade is elevated not by pneumatics or linear actuators, both of which would give it the wrong look, but by a PF M motor and a clever system of rotating pushers at the chassis' bottom. The brick-built body is teeming with details and is complemented by custom stickers.

## CHALLENGES

Other than fitting so many functions in a relatively small model without compromising its looks, the primary challenge was the lights. The original D9T has 11 lights, and using LEGO LEDs would make the model larger and uglier. The model uses smaller, brighter, third-party LEDs inserted into 1×1 bricks with a custom plug that connects them to the Power Functions system.

*The cabins of most large dozers are fitted with bulletproof glass to protect the driver during demolition operations. There are also completely armored dozers used by the military for salvage operations in combat conditions.*

## THE ORIGINAL

At almost 50 tons, the D9T is one of the largest Caterpillar dozers. This reliable machine is well suited for quarries, forestry, landfills, mine sites, and construction and demolition sites. It features Caterpillar's distinctive triangular tracks with the topmost sprockets driven, which allows extra clearance between the ground and its transmission.

**+ FUN FACT**

*The model is able to operate in winter weather at 5°F (−15°C). The extra ground clearance from triangular tracks comes in handy for driving through snow!*

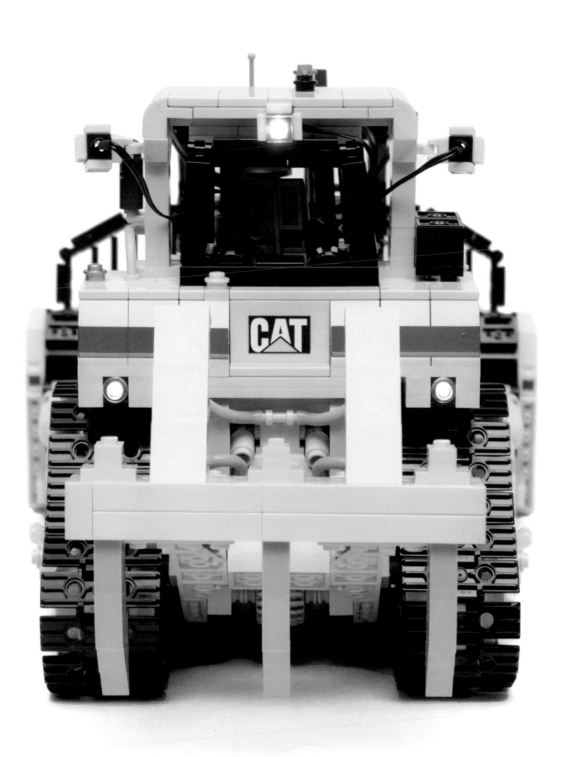

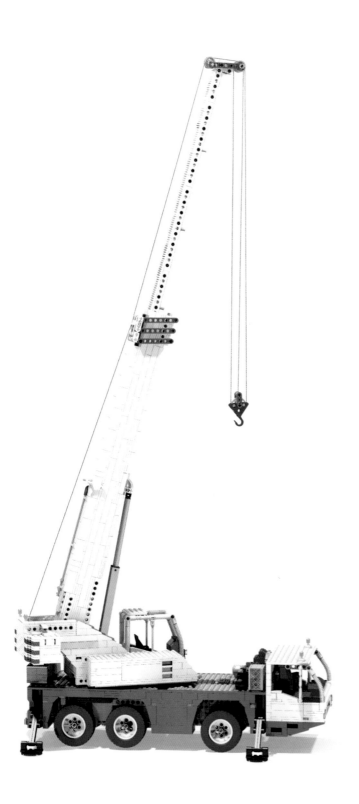

# DEMAG AC50-1

*Jennifer Clark (2002)*

## ABOUT THE MODEL

Built years before the introduction of the Power Functions system, this mobile crane model amazes with its complexity and incredible look. The carrier is fitted with a suspension, Ackermann steering geometry, two driven axles, and motorized outriggers, while the rotating superstructure includes a motorized winch and a massive telescopic boom that can be elevated and extended.

The exterior is expertly crafted with LEGO bricks, and the livery and custom stickers are based on vehicles operated by Baldwins Crane Hire Ltd.

## CHALLENGES

This model does a fantastic job overcoming typical challenges of the early Technic era: the lack of strong motors and linear actuators, the modest selection of studless pieces, and the simple electric system with no remote-control option. A Firgelli linear actuator is used to handle the boom, and the initial third-party radio control system was later replaced with LEGO PF IR receivers. All functions are operated by 71427 9V motors, and the chassis that houses all this functionality is, amazingly, primarily made of the old-style Technic bricks, rather than the modern beams.

## SPECIFICATIONS

| | |
|---|---|
| LENGTH | **18.8"** |
| WIDTH | **10.1"** |
| HEIGHT | **39.5"** |
| PIECES | **3,362** |

## THE ORIGINAL

Manufactured by the German Demag company, now owned by Terex Corporation, the AC50-1 is an exemplary small modern crane designed with effectiveness in mind. All-wheel drive and an advanced suspension system make it a real all-terrain crane, while speed-dependent rear-axle steering adds the extra maneuverability needed for an urban environment. With a load capacity of 55 tons and a boom that extends up to 190 feet (58 meters), the AC50-1 enjoys a reputation for reliability and well-deserved popularity.

**+ FUN FACT**

*This model was one of the first models that pushed LEGO Technic to its absolute limits, and it went viral in the early 2000s. The model sparked a community of custom builders, who attempted to create something even half as complex. It's credited with attracting countless enthusiasts to LEGO Technic.*

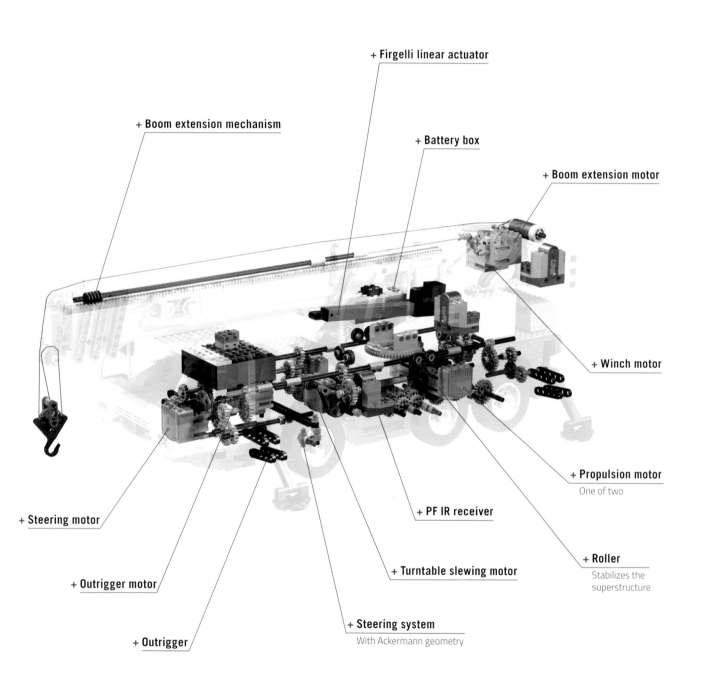

+ Firgelli linear actuator

+ Boom extension mechanism

+ Battery box

+ Boom extension motor

+ Winch motor

+ Steering motor

+ Propulsion motor
One of two

+ PF IR receiver

+ Outrigger motor

+ Roller
Stabilizes the
superstructure

+ Turntable slewing motor

+ Steering system
With Ackermann geometry

+ Outrigger

# JCB JS220

*Jennifer Clark (2002)*

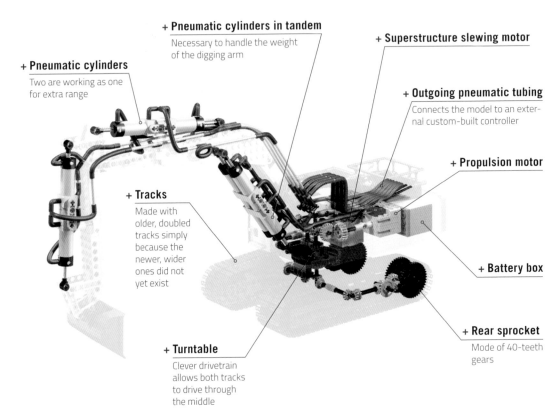

**+ Pneumatic cylinders in tandem**
Necessary to handle the weight of the digging arm

**+ Superstructure slewing motor**

**+ Pneumatic cylinders**
Two are working as one for extra range

**+ Outgoing pneumatic tubing**
Connects the model to an external custom-built controller

**+ Propulsion motor**

**+ Tracks**
Made with older, doubled tracks simply because the newer, wider ones did not yet exist

**+ Battery box**

**+ Rear sprocket**
Mode of 40-teeth gears

**+ Turntable**
Clever drivetrain allows both tracks to drive through the middle

## ABOUT THE MODEL

This model of the JCB excavator features motorized tracks and superstructure rotation, as well as a pneumatically operated digging arm. It's extremely realistic not only in outward appearance but also in internal design, with all three motors located in the superstructure and both tracks driven through the turntable by a brilliant drivetrain.

The smooth body is teeming with details and covered in custom stickers based on the real machine. Wires and pneumatic tubes connect the model to a custom-built controller, which includes pneumatic valves and a motorized compressor with a pressure safety switch.

## CHALLENGES

As with all early Technic models, it was challenging to power it with weak 9V motors and to build a strong and compact structure with mostly classic LEGO pieces, but the main challenge was the arm. It needed doubled pneumatic cylinders to achieve a realistic movement range, and it had to be heavily reinforced to allow actual digging.

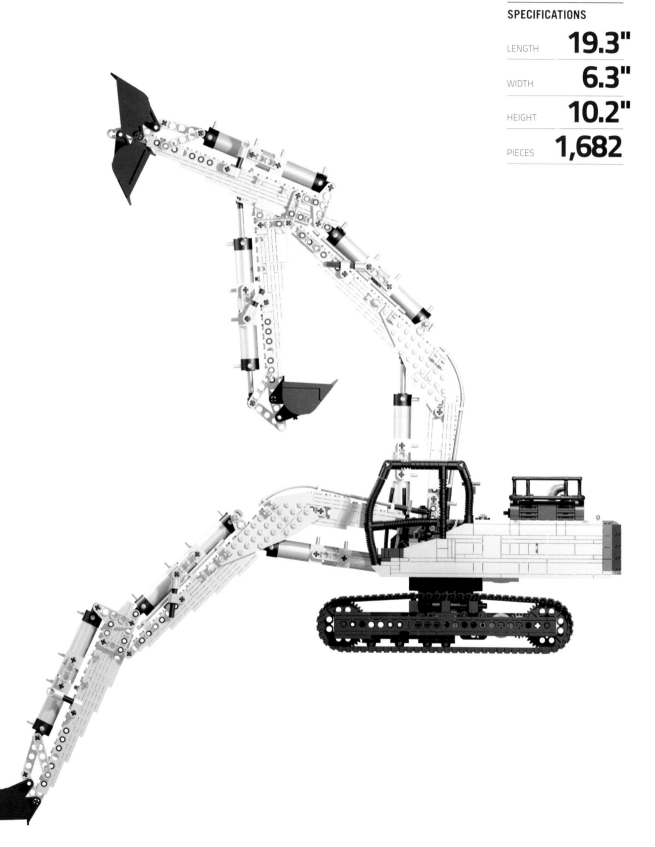

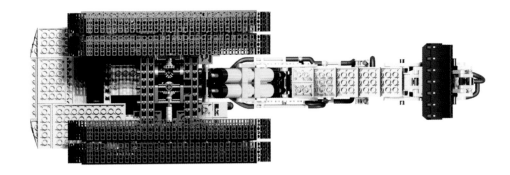

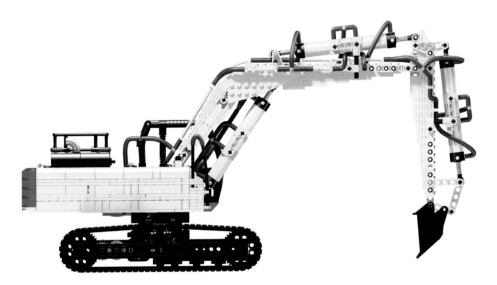

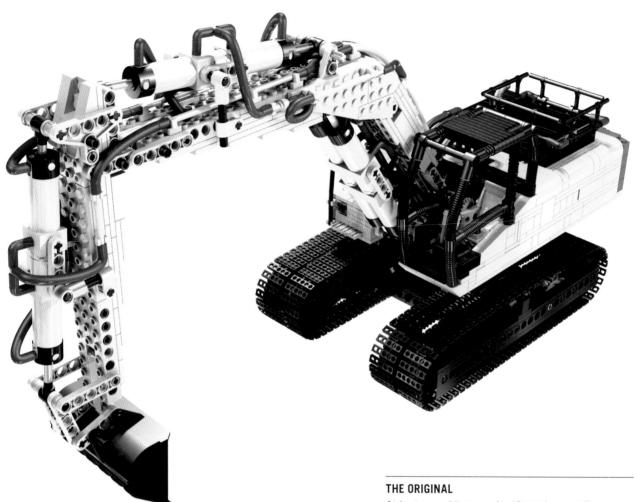

### THE ORIGINAL

At just over 22 tons, the JS220 is a medium-sized machine in the JCB excavators range. Benefiting from nearly 50 years of the company's experience, the vehicle offers a number of improvements in efficiency, comfort, and safety over the previous generations. It comes with four preset working modes, an advanced hydraulic drive that can recycle oil, and pipework that can be customized to better suit its working conditions.

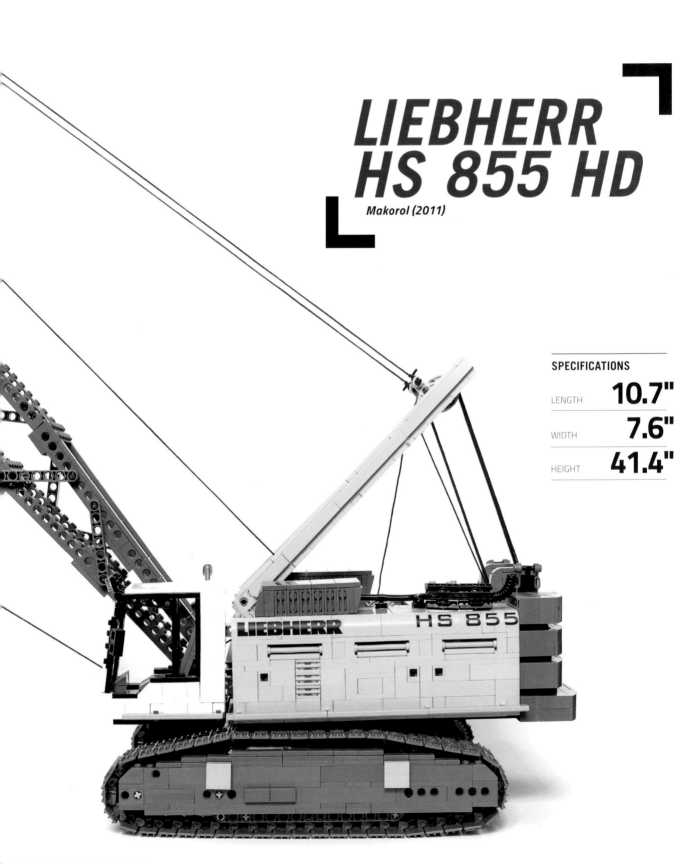

# LIEBHERR HS 855 HD

*Makorol (2011)*

## SPECIFICATIONS

LENGTH **10.7"**

WIDTH **7.6"**

HEIGHT **41.4"**

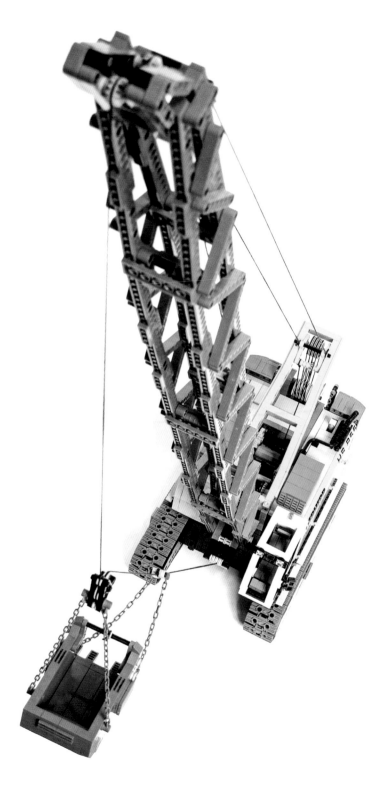

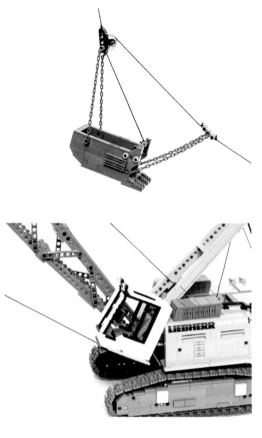

## ABOUT THE MODEL

This model of a Liebherr dragline excavator is split into two independent parts. The chassis has its own power supply and IR receiver, which means the superstructure can make an unlimited number of rotations, because not a single wire connects it to the chassis.

The model is propelled by two PF M motors, while five more motors are slewing the superstructure, elevating the cabin, and driving the winches. The body is entirely built using bricks, with plenty of details and a carefully re-created Liebherr livery. The boom combines bricks and beams to create a sturdy truss structure, and the counterweight houses a number of weighted bricks to act like the real machine. The bucket is carefully shaped with small pieces.

## CHALLENGES

The main difficulty was understanding how the real machine operates and modeling it with LEGO pieces. The bucket of a real dragline excavator performs a complex motion, which is affected by the tension and lengths of two lines, as well as the length and attachment points of chains. It took a lot of trial and error to reverse engineer this functionality, and then it turned out that actually controlling the model took some practice, too!

## THE ORIGINAL

The Liebherr HS 855 is essentially a regular crawler crane. Propelled by a 612 hp engine, this 84-ton machine develops speeds up to 0.83 mph (1.34 km/h). Vastly popular, it can be adapted to a whole array of applications, effectively acting as a standard crane, a demolition crane, a pipelayer, a drilling rig, a clamshell or dragline excavator, and more.

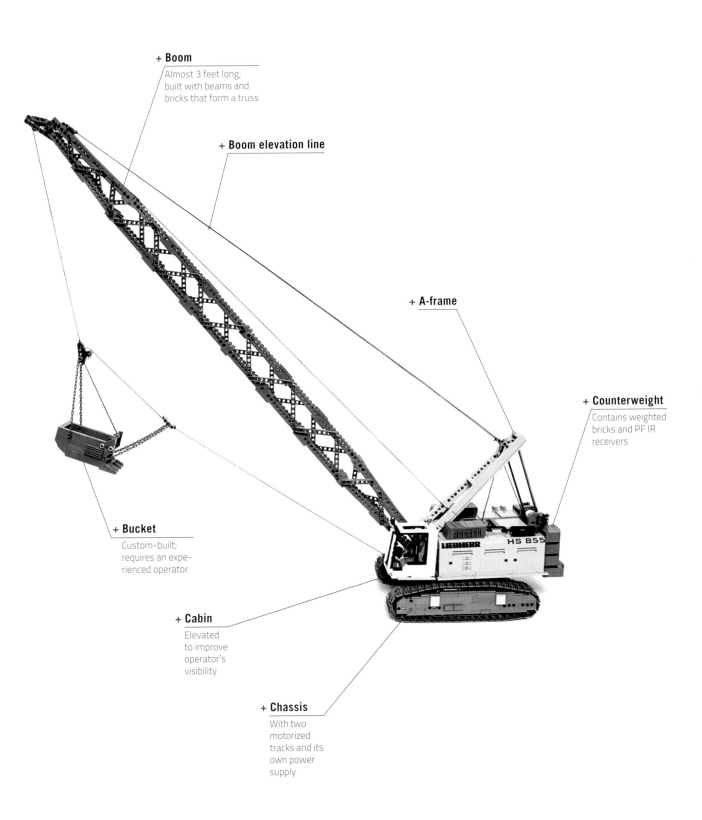

**+ Boom**
Almost 3 feet long;
built with beams and
bricks that form a truss

**+ Boom elevation line**

**+ A-frame**

**+ Counterweight**
Contains weighted
bricks and PF IR
receivers

**+ Bucket**
Custom-built;
requires an expe-
rienced operator

**+ Cabin**
Elevated
to improve
operator's
visibility

**+ Chassis**
With two
motorized
tracks and its
own power
supply

LIEBHERR    HS 855

# LIEBHERR L 580

*M_Longer (2009)*

## SPECIFICATIONS

| | |
|---|---|
| LENGTH | **20.5"** |
| WIDTH | **11.3"** |
| HEIGHT | **6.3"** |
| PIECES | **1,604** |

## ABOUT THE MODEL

This model of the Liebherr front-end loader comes with a 4x4 drive, articulated steering, and a fully motorized arm. All functions are remote controlled, the rear axle has a pendular suspension, and the headlights and taillights are fitted with LEGO LEDs. The body is mostly built using bricks, with colors and plenty of details taken from the original L 580, such as the fully glazed cabin with an opening door and complete interior.

## CHALLENGES

The arm! Its movement range had to be accurately re-created to let it operate smoothly, and the bucket had to be reasonably light. In fact, this model uses a LEGO ready-made bucket that is slightly too small for it because a custom-built bucket at scale proved too heavy.

## THE ORIGINAL

The Liebherr L 580 is a large 25-ton-class front-end loader, built with a focus on the operator's comfort and equipped with advanced efficiency-monitoring software. The 292 hp engine and 4x4 drive make it suitable for a variety of applications beyond the usual material handling, such as demolition and forestry.

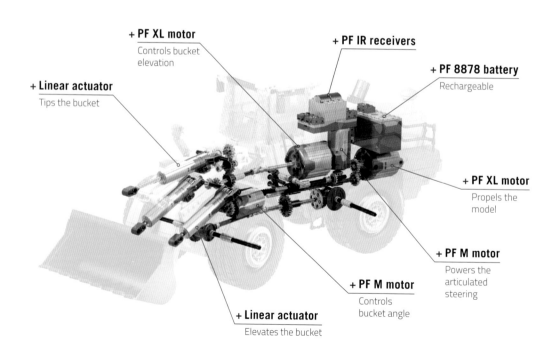

**+ PF XL motor**
Controls bucket elevation

**+ PF IR receivers**

**+ PF 8878 battery**
Rechargeable

**+ Linear actuator**
Tips the bucket

**+ PF XL motor**
Propels the model

**+ PF M motor**
Powers the articulated steering

**+ PF M motor**
Controls bucket angle

**+ Linear actuator**
Elevates the bucket

**+ FUN FACT**
*This loader was meant as a simple model just to accompany a dump truck model, but the builder got carried away.*

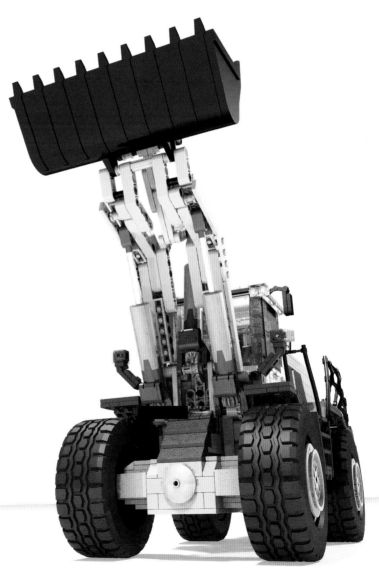

# LIEBHERR
# LTM 1050-3.1

*Makorol (2010)*

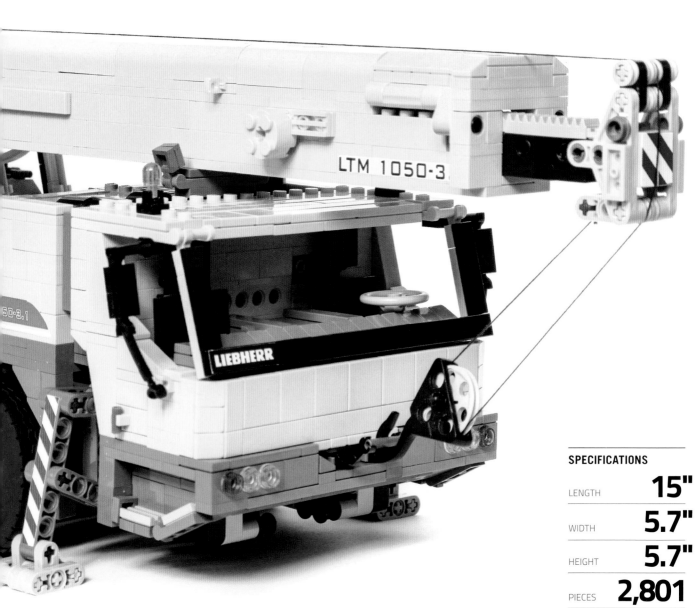

**SPECIFICATIONS**

| | |
|---|---|
| LENGTH | **15"** |
| WIDTH | **5.7"** |
| HEIGHT | **5.7"** |
| PIECES | **2,801** |

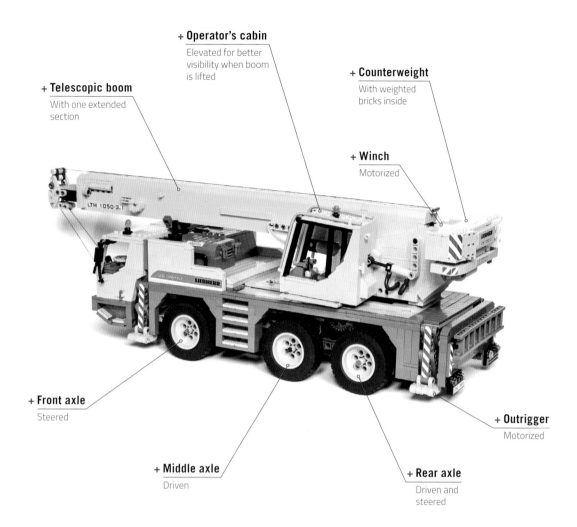

**+ Telescopic boom**
With one extended section

**+ Operator's cabin**
Elevated for better visibility when boom is lifted

**+ Counterweight**
With weighted bricks inside

**+ Winch**
Motorized

**+ Front axle**
Steered

**+ Middle axle**
Driven

**+ Rear axle**
Driven and steered

**+ Outrigger**
Motorized

LTM 1050-3.1

LIEBHERR

**ABOUT THE MODEL**

Mobile cranes are a popular yet challenging subject for the Technic builders. This one stands out by blending great functionality with superb looks at a medium scale. It's powered by seven motors and a pneumatic system, with the two last axles driven and the first and last axles steered. The four outriggers, the winch, and the boom extension mechanism are all motorized, and the superstructure is slewed by a motor as well. Another motor powers a pneumatic compressor, which feeds cylinders that lift the boom and elevate the operator's cabin. Even the counterweight has real weighted bricks inside it!

**THE ORIGINAL**

The Liebherr LTM 1050 is a modern, medium-sized mobile crane. It's fully computerized, propelled by a 367 hp diesel engine through a 12-speed automatic transmission, fitted with hydro-pneumatic suspension, and capable of lifting up to 50 tons of load while weighing just 36 tons itself. Great maneuverability, resulting from all-wheel steering, and the overall ease of operation make the LTM 1050-3.1 a popular choice for a variety of applications.

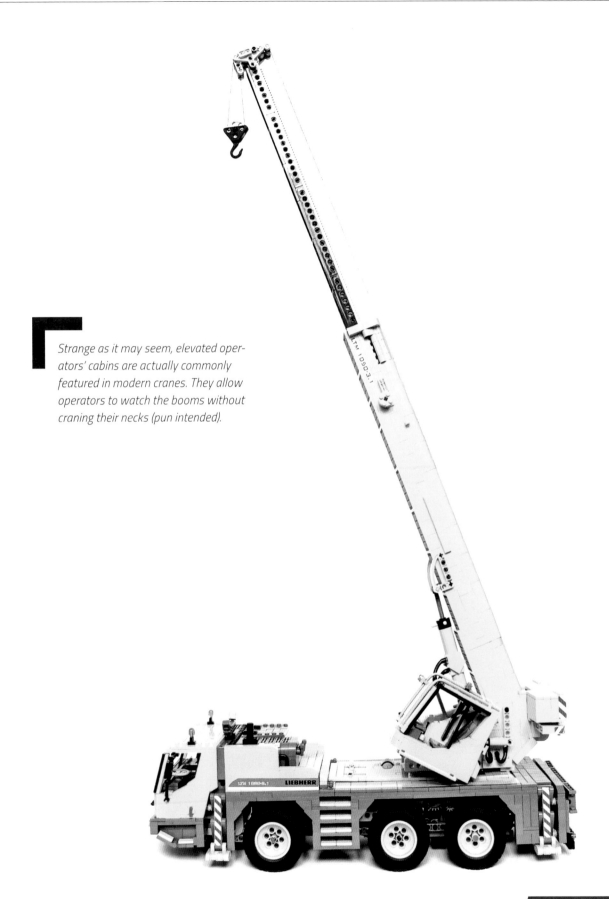

Strange as it may seem, elevated operators' cabins are actually commonly
featured in modern cranes. They allow
operators to watch the booms without
craning their necks (pun intended).

# LIEBHERR PR 764 LITRONIC

*M_Longer (2010)*

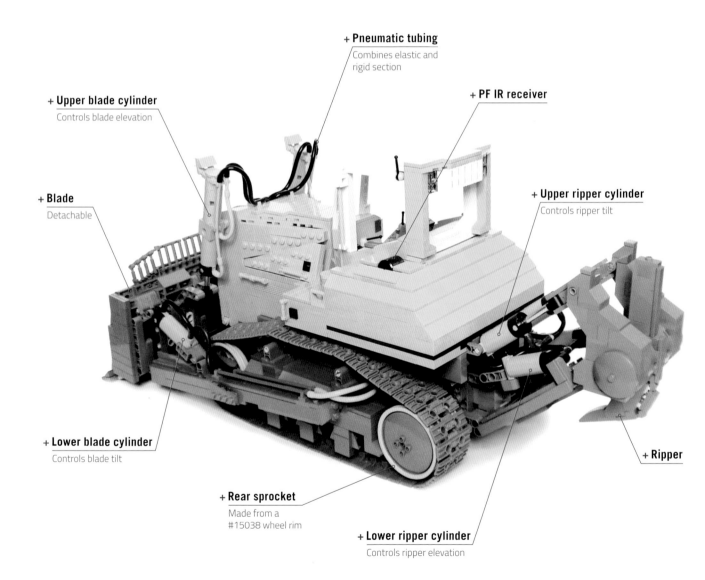

+ **Pneumatic tubing**
Combines elastic and rigid section

+ **PF IR receiver**

+ **Upper blade cylinder**
Controls blade elevation

+ **Upper ripper cylinder**
Controls ripper tilt

+ **Blade**
Detachable

+ **Lower blade cylinder**
Controls blade tilt

+ **Ripper**

+ **Rear sprocket**
Made from a
#15038 wheel rim

+ **Lower ripper cylinder**
Controls ripper elevation

LENGTH **19.7"**

WIDTH **9.4"**

HEIGHT **7.9"**

PIECES **~2,500**

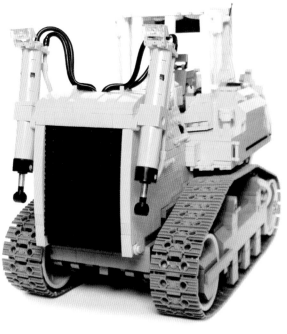

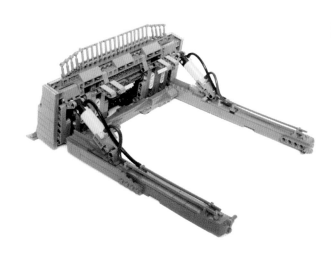

### ABOUT THE MODEL

This model of Liebherr's largest bulldozer features motorized tracks and a pneumatically operated front blade and rear ripper that can be elevated and tilted. It's fully remote controlled, with eight pneumatic cylinders fed by an internal compressor, four pneumatic valves, and a safety switch that turns the compressor off automatically before the air pressure becomes critical.

The body is carefully shaped using LEGO slopes and bricks, matching the classic Liebherr livery and concealing most of the mechanic and electric elements.

### CHALLENGES

Making a pneumatic system this complex and fully remote controlled was difficult and space consuming. The compressor and its safety switch fill most of the space under the hood, while the complicated system of motorized valves is located below the cabin. Both the blade and the ripper have two cylinders for elevation and another two for tilting, requiring four separate pneumatic circuits. All this effort ensures the smooth action of the pneumatic system.

### THE ORIGINAL

PR 764 is the largest bulldozer produced by Liebherr. At a weight of more than 50 tons and with its 422 hp engine, the PR 764 is a heavy-duty machine popular at construction sites and in demolition, forestry, and waste handling. A hydraulic drive makes sure it has enormous torque, while its top speed is just 7 mph (11 km/h).

# SANDVIK LH 517L

*M_Longer (2010)*

## ABOUT THE MODEL

This model of an underground Sandvik front loader is fitted with all-wheel drive, articulated steering, and an enormous custom-built bucket that is operated pneumatically. Designed to work in mines, it has an array of working lights, even inside the cabin. The rear axle has suspension, and the body is meticulously re-created with custom stickers. The model is extremely accurate because the builder sees the original machine on a daily basis.

## CHALLENGES

The model is actually rather small; its size is limited by wheels, which are simply the largest ones LEGO produces. With plenty of internal space taken by rear suspension, linear actuators were too large for this model. Instead, the pneumatic system that controls the bucket is controlled from the outside, by means of pneumatic tubing that forms a "leash" and connects the model to external valves and pumps. The drive and steering are fully remote controlled.

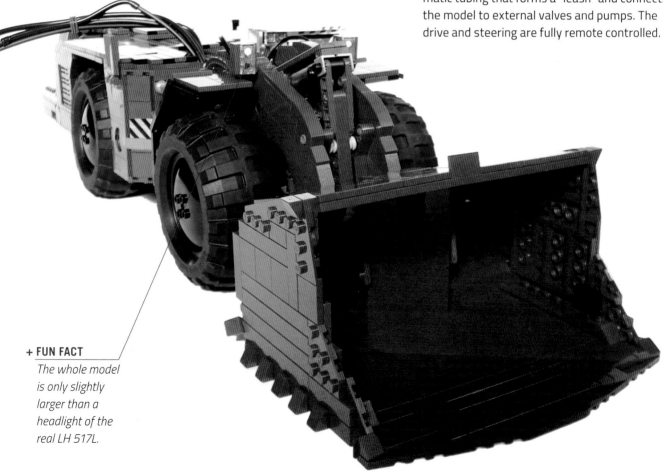

**+ FUN FACT**

*The whole model is only slightly larger than a headlight of the real LH 517L.*

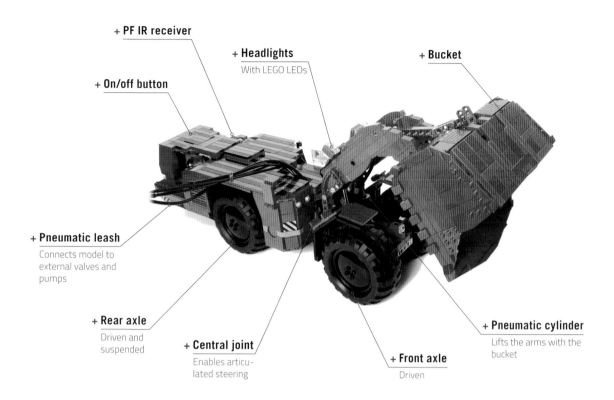

+ **PF IR receiver**

+ **Headlights**
With LEGO LEDs

+ **Bucket**

+ **On/off button**

+ **Pneumatic leash**
Connects model to
external valves and
pumps

+ **Rear axle**
Driven and
suspended

+ **Central joint**
Enables articu-
lated steering

+ **Front axle**
Driven

+ **Pneumatic cylinder**
Lifts the arms with the
bucket

*The black strips on top of the
hull aren't paint but sand-
paper. They prevent maintenance workers from slipping
while servicing the engine.*

## SPECIFICATIONS

| | |
|---|---|
| LENGTH | **23.6"** |
| WIDTH | **6.7"** |
| HEIGHT | **4.7"** |
| PIECES | **~1,800** |

## THE ORIGINAL

The Sandvik company has more than 150 years
of tradition in manufacturing equipment for
mining, much of it made specifically to operate underground, like the LH 517L front loader.
This 36-foot (11-meter) long colossus weighs
44 tons and is fitted with wheels taller than a
person. Propelled by a 388 hp diesel engine, it's
well suited to moving through mine tunnels due
to its narrow, low body and powerful spotlights.

# SANDVIK PF300

*Konajra (2011)*

The PF300 is such a new design that the LEGO model was finished before the real machine's first unit was actually assembled.

## SPECIFICATIONS

LENGTH **32.3"**

WIDTH **15"**

HEIGHT **10.5"**

PIECES **~9,200**

### ABOUT THE MODEL

This model of the massive, mobile crushing plant for strip mines was created to accompany the builder's excavator models, such as the Caterpillar 7495 HF on page 98. Built at LEGO minifig scale, the crusher features four moving tracks, with a steering system on two, and a motorized conveyor belt. With a massive, carefully shaped chute and a classic LEGO body that is full of details such as catwalks, railings, lamps, and engines, the model is so accurate that the Sandvik company keeps it on permanent display at its headquarters.

### CHALLENGES

The biggest challenge was the structural integrity of a model that is heavy and tall and rests on a thin V-shaped pillar on a narrow chassis. The model has no motorized propulsion system because it could make the model collapse.

### THE ORIGINAL

Designed for strip mines, a crushing plant steps in between excavators and output depots, where dump trucks are traditionally used. The crushing plants can collect and pre-process output directly from excavators and then send it onward on a loading bridge or a beltwagon. Using a crushing plant is cheaper, more effective, and more environmentally friendly than using a fleet of trucks. Built by the Swedish Sandvik company, which specializes in high-tech mining equipment, the PF300 is distinguished by its compactness, high capacity, and ability to work with various materials. It's also highly mobile and can follow excavators to where it's most needed.

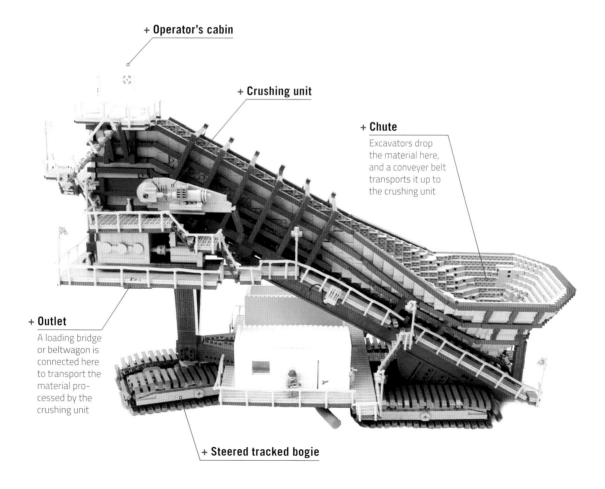

+ **Operator's cabin**

+ **Crushing unit**

+ **Chute**

Excavators drop the material here, and a conveyer belt transports it up to the crushing unit

+ **Outlet**

A loading bridge or beltwagon is connected here to transport the material processed by the crushing unit

+ **Steered tracked bogie**

# ZOREX EXCAVATOR

*Jurgen Krooshoop (2012)*

## SPECIFICATIONS

LENGTH **17.4"**

WIDTH **7.9"**

HEIGHT **8.8"**

PIECES **1,688**

## ABOUT THE MODEL

Inspired by both Hyundai excavators and the LEGO Motorized Excavator set (#8043), this model features drive, steering, slewing, and a three-section arm, all of which are motorized and can operate at the same time. The arm's working range is realistic, and the tracks have a suspension system. Two versions of this model exist: the Zorex-220 (orange, created in 2011) and the newer and improved Link Belt 250 X 3 (white/red, created in 2014).

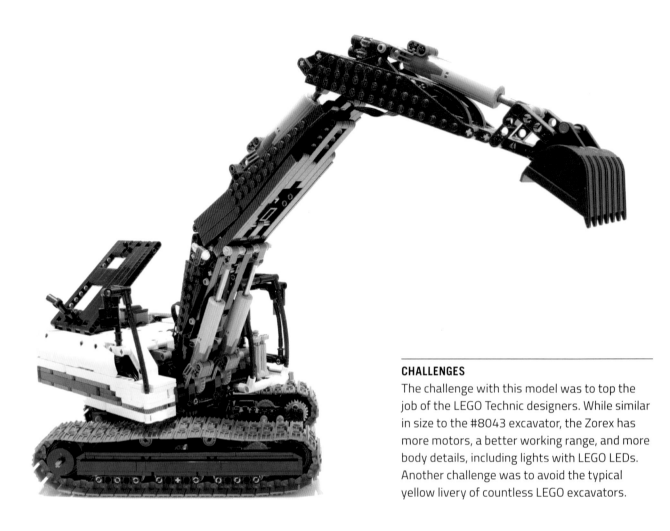

## CHALLENGES

The challenge with this model was to top the job of the LEGO Technic designers. While similar in size to the #8043 excavator, the Zorex has more motors, a better working range, and more body details, including lights with LEGO LEDs. Another challenge was to avoid the typical yellow livery of countless LEGO excavators.

# MISCELLANEOUS

# *BRAIDING MACHINE*

*Nico71 (2013)*

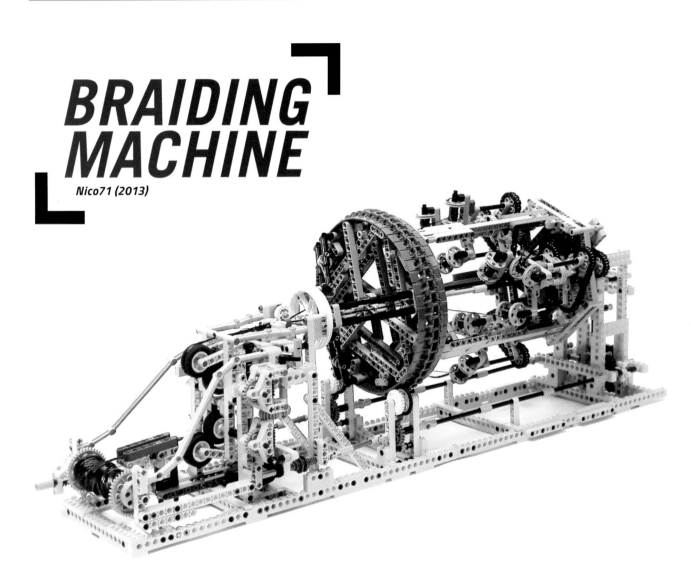

## ABOUT THE MODEL

This unique creation is where LEGO meets fashion. The Braiding Machine can produce a 5-foot (1.5-meter) long wristband out of nine strings in a little over four minutes—and all using just a single motor. The machine, which took nine months to build, uses an incredibly complex system of gear wheels, chains, differentials, and turntables to weave a single string using three sets of planetary reels. A section of Technic track forms a loop, acting as an extra-large gear wheel. The entire device takes 15 minutes to set up, and it can work unattended. The end result is a woven wristband that the model pulls out and stores on an output drum. Just come back to pick it up!

## CHALLENGES

Using a single motor to power the entire device required dozens of moving parts to be connected, synchronized, and rotating in the right direction at the right speed, while producing as little friction as possible. At the same time, the weaving process needs the band to be tightened and pulled at a specific rate and carefully led to prevent it from coming off the many rollers and drums. Plenty of effort was devoted to balance the tension of all the strings against the rigidity of the machine.

## SPECIFICATIONS

| | |
|---|---|
| LENGTH | **29.2"** |
| WIDTH | **7.9"** |
| HEIGHT | **10"** |
| PIECES | **1,929** |

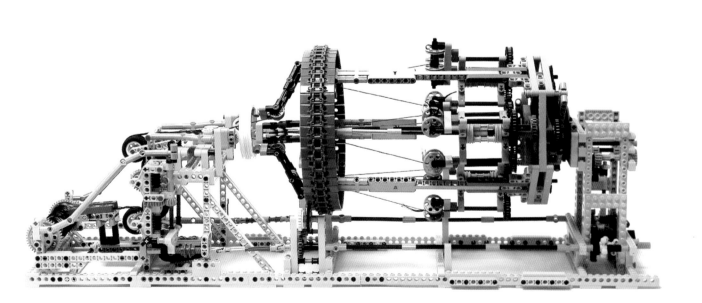

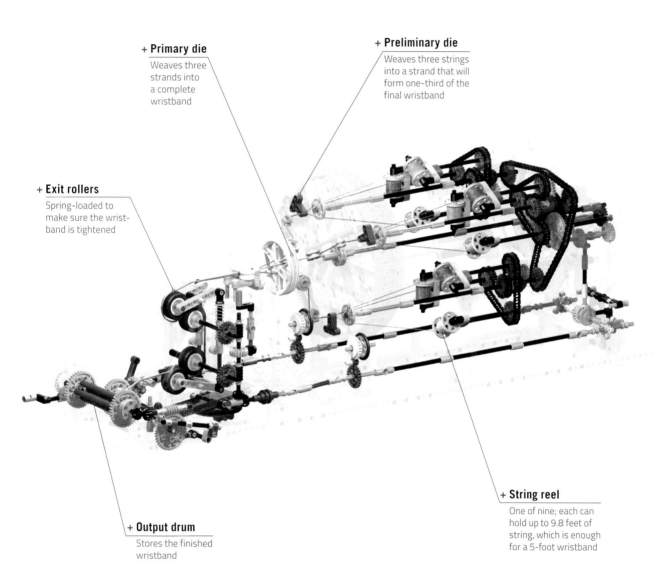

**+ Primary die**
Weaves three
strands into
a complete
wristband

**+ Preliminary die**
Weaves three strings
into a strand that will
form one-third of the
final wristband

**+ Exit rollers**
Spring-loaded to
make sure the wrist-
band is tightened

**+ Output drum**
Stores the finished
wristband

**+ String reel**
One of nine; each can
hold up to 9.8 feet of
string, which is enough
for a 5-foot wristband

# DA VINCI FLYER

*Mahj (2010)*

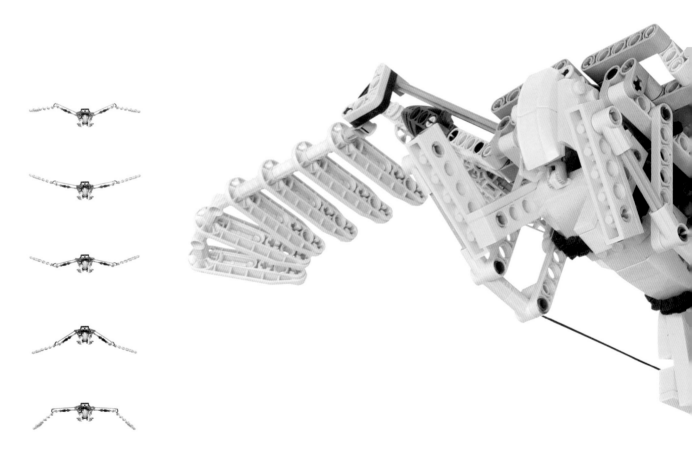

+ The wings are shaped
  with extensive use of
  LEGO Bionicle pieces.

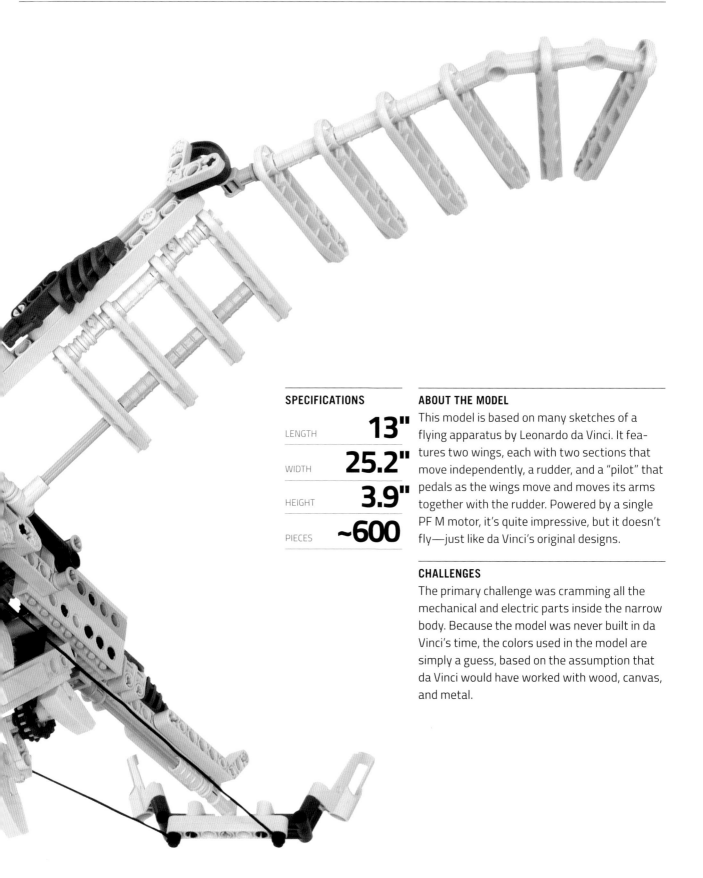

## SPECIFICATIONS

| | |
|---|---|
| LENGTH | **13"** |
| WIDTH | **25.2"** |
| HEIGHT | **3.9"** |
| PIECES | **~600** |

## ABOUT THE MODEL

This model is based on many sketches of a flying apparatus by Leonardo da Vinci. It features two wings, each with two sections that move independently, a rudder, and a "pilot" that pedals as the wings move and moves its arms together with the rudder. Powered by a single PF M motor, it's quite impressive, but it doesn't fly—just like da Vinci's original designs.

## CHALLENGES

The primary challenge was cramming all the mechanical and electric parts inside the narrow body. Because the model was never built in da Vinci's time, the colors used in the model are simply a guess, based on the assumption that da Vinci would have worked with wood, canvas, and metal.

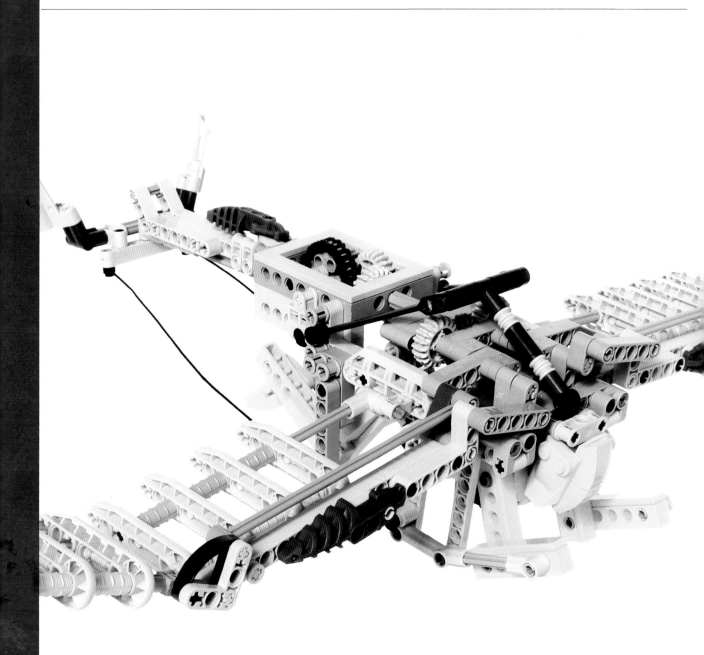

Da Vinci had a habit of writing all his notes in "mirror handwriting." It's difficult to know what his reason was, but he may have been trying to protect his documents from unauthorized readers.

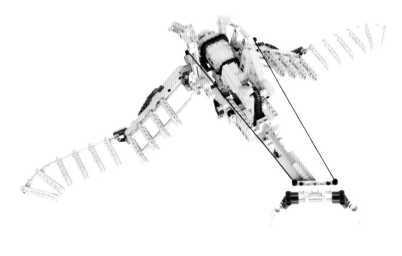

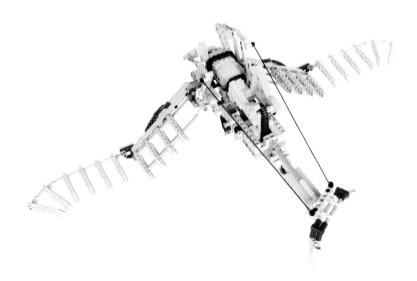

### THE ORIGINAL

Da Vinci dreamed of an apparatus that would make human flight possible, and he studied birds extensively, believing that re-creating their body mechanics was the key to flight. His work on emulating the natural movements of a bird's wings was advanced, but there are no records of any of his designs ever being built. While mechanically brilliant, da Vinci's apparatus would never take off because of the weight of the pilot.

# PEGASUS AUTOMATON

*Amida (2013)*

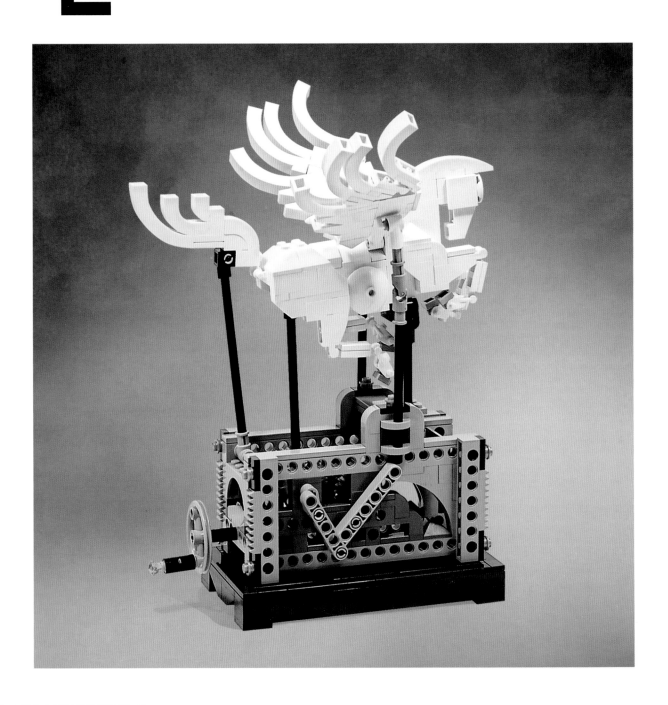

## SPECIFICATIONS

LENGTH **7.5"**

WIDTH **6.4"**

HEIGHT **10.2"**

PIECES **480**

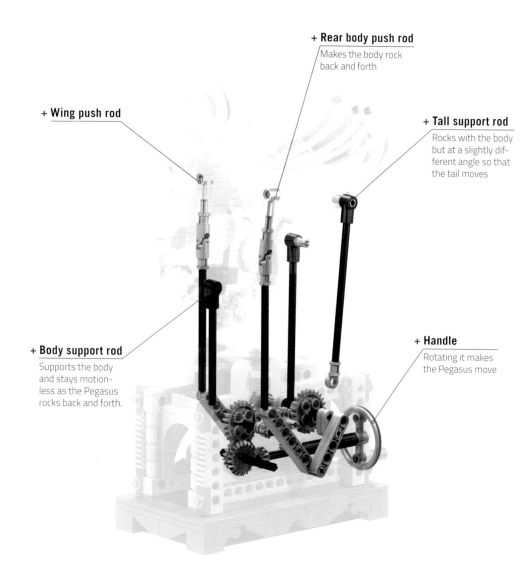

**+ Wing push rod**

**+ Rear body push rod**
Makes the body rock back and forth

**+ Tall support rod**
Rocks with the body but at a slightly different angle so that the tail moves

**+ Body support rod**
Supports the body and stays motionless as the Pegasus rocks back and forth.

**+ Handle**
Rotating it makes the Pegasus move

### ABOUT THE MODEL

*Automata*, or kinetic sculptures, are self-operated mechanisms capable of executing complex movements that originated in ancient Greece. They were often built to mimic animal or human motion, as well as demonstrate the mechanical skill of their builder. Most historical automata represent animals, real or mythological, such as this Pegasus Automaton. It is operated by a clever system of gears and cranks that make it beat its wings and move its tail up and down as you crank its handle.

### CHALLENGES

The combined weight of the moving parts, especially the wings, is surprisingly heavy. The model required a complicated set of gears to achieve a sufficiently high gear reduction within a confined space.

# TACHIKOMA

*Mahj (2012)*

## SPECIFICATIONS

| | |
|---|---|
| LENGTH | **9.8"** |
| WIDTH | **8.7"** |
| HEIGHT | **8.7"** |
| PIECES | **950** |

## ABOUT THE MODEL

This model of the Tachikoma robot from **Ghost in the Shell** uses a complex chassis that allows driving and walking at the same time. Each of the four legs is tipped with a small wheel, driven by one of two PF M motors, which makes skid steering possible. A PF XL motor powers the walking mechanism that moves all the legs.

## CHALLENGES

The chassis design, which needed to be compact and stable despite the robot's tall silhouette, proved extremely difficult. It took trial and error to get the legs to move in a way that would allow the robot to advance without toppling over or swaying in place.

## THE ORIGINAL

Tachikomas are "think tanks" equipped with artificial intelligence. They use optical camouflage and can drive, climb, and jump, and they accommodate a single passenger.

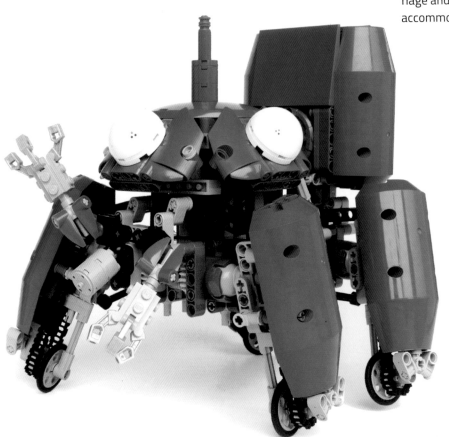

*Masamune Shirow, the creator of **Ghost in the Shell**, based his robots on jumping spiders (Salticidae). You can see how these spiders inspired the many eyes, strong legs, and arachnid body shape.*

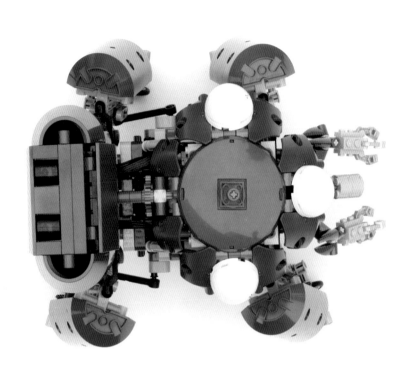

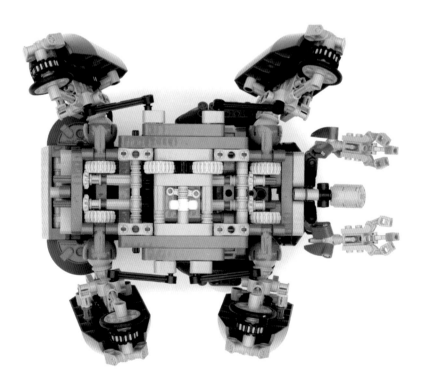

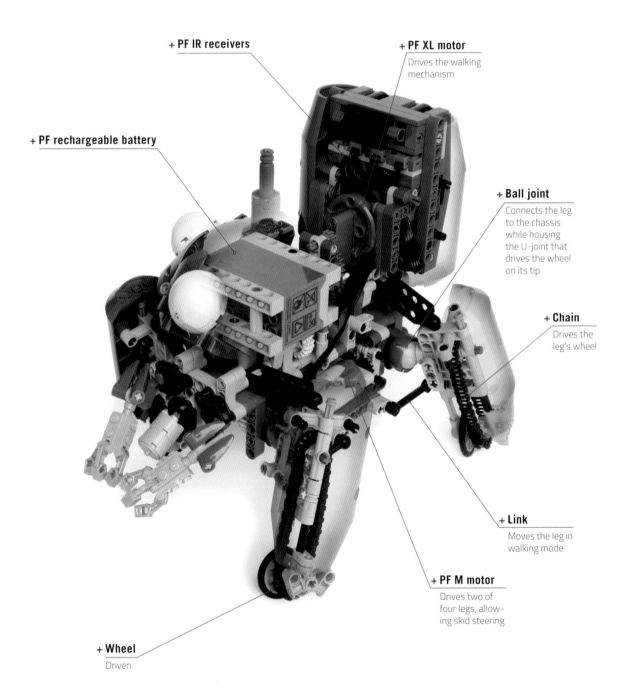

**+ PF IR receivers**

**+ PF XL motor**
Drives the walking
mechanism

**+ PF rechargeable battery**

**+ Ball joint**
Connects the leg
to the chassis
while housing
the U-joint that
drives the wheel
on its tip

**+ Chain**
Drives the
leg's wheel

**+ Link**
Moves the leg in
walking mode

**+ PF M motor**
Drives two of
four legs, allow-
ing skid steering

**+ Wheel**
Driven

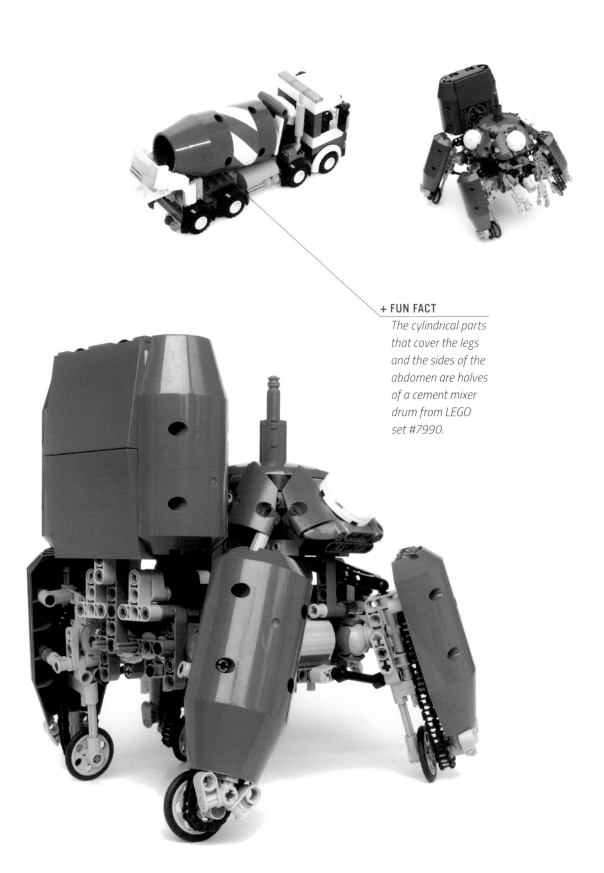

**+ FUN FACT**
*The cylindrical parts that cover the legs and the sides of the abdomen are halves of a cement mixer drum from LEGO set #7990.*

# TEKNOMEKA

*Klaupacius (2005)*

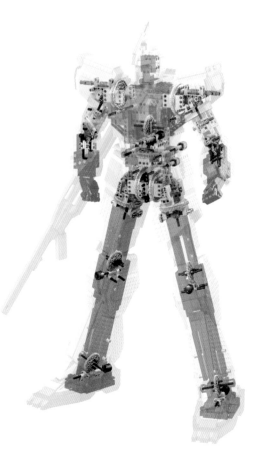

### ABOUT THE MODEL

You can't help but feel pure joy when you see a LEGO mecha model standing more than 2 feet (0.6 meters) tall. Teknomeka is the ultimate big robot toy you've always wanted. The entire model is built around a complex Technic skeleton with 20 articulated joints—each one can be locked using worm gears. The construction is modular, which allows for easy repairs and transportation; the outer skin is removable; and the arms include breakaway joints that protect them in case of collapse.

In keeping with tradition, the Teknomeka comes with a chest compartment that houses a single LEGO minifigure; this is the robot's pilot. The tiny pilot showcases the model's massive size. There are no plans for a hamster-piloted version as yet.

## SPECIFICATIONS

| | |
|---|---|
| LENGTH | **8.5"** |
| WIDTH | **16.3"** |
| HEIGHT | **28.7"** |
| PIECES | **4,180** |

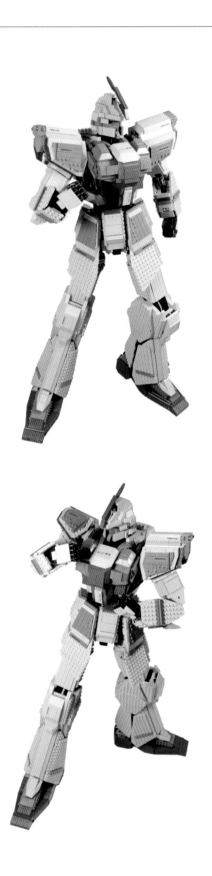

Because the builder of
this model has created
detailed, step-by-step
building instructions,
fans have used his work
as a starting point for
their own giant anime-
inspired robots.

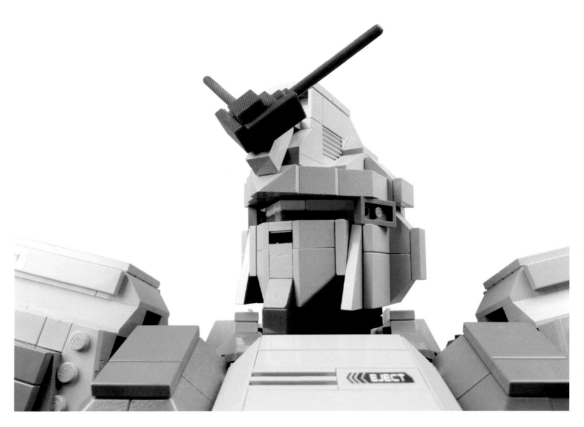

## CHALLENGES

The chief challenge with a tall LEGO robot is balance. The model needs to be able to stand on its own in various poses and with large loads acting on its joints for prolonged periods of time. The Teknomeka model's skeletal structure provides easily accessible knobs for fine adjustments to the angle of the ankles, hips, and waist in order to achieve static balance.

# MOTORCYCLES

| | |
|---|---|
| LENGTH | **12.6"** |
| WIDTH | **3.9"** |
| HEIGHT | **7.2"** |
| PIECES | **~900** |

# HONDA CBR1000RR REPSOL

陳彥璋 *[Oryx Chen] (2012)*

## ABOUT THE MODEL

While all Technic motorbikes are difficult to build, this model ups the ante with its beautiful custom stickers. Its functions include a full suspension (with Honda's unique pro-link rear suspension), a working piston engine connected to the rear wheel by a chain, a folding side stand and rear footrests, and a small trunk under the backseat. The exterior was re-created primarily with non-Technic pieces, most of which are slopes, and the entire model is robust enough to be handled roughly. The rims are custom painted, and electric wires are used as brake cables.

## CHALLENGES

The primary challenge was the small scale, which is limited by the size of LEGO wheels. At this scale, certain shapes could only be approximated. It was also difficult to create a strong but compact frame to hold the model together while leaving enough space for its functions to work. Finally, the custom stickers were meticulously cut to fit specific LEGO elements.

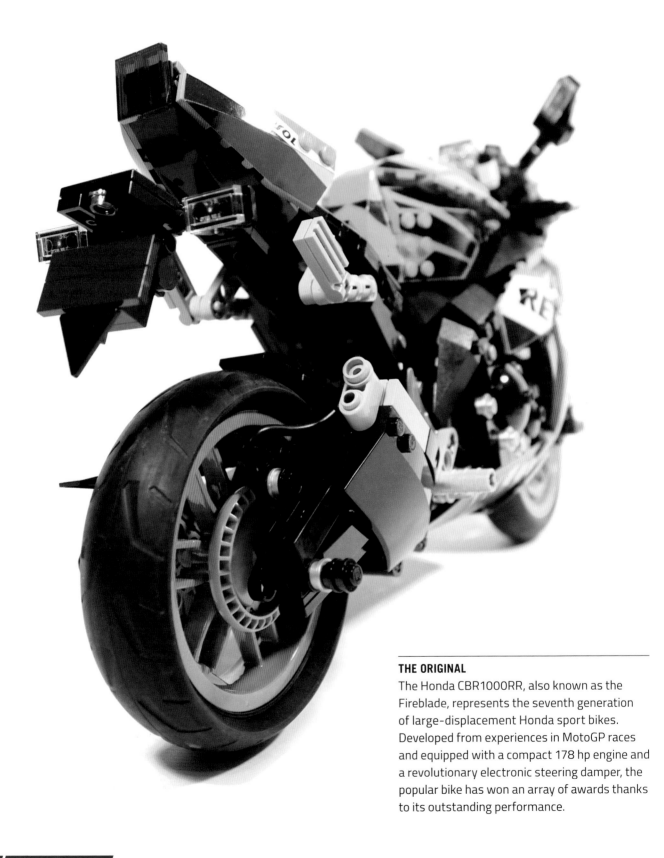

### THE ORIGINAL

The Honda CBR1000RR, also known as the Fireblade, represents the seventh generation of large-displacement Honda sport bikes. Developed from experiences in MotoGP races and equipped with a compact 178 hp engine and a revolutionary electronic steering damper, the popular bike has won an array of awards thanks to its outstanding performance.

The Honda factory racing team started working with the Spanish Repsol company nearly 20 years ago and since then has become the most successful superbike racing team ever, with 11 world titles, 124 individual wins, and 338 podium finishes.

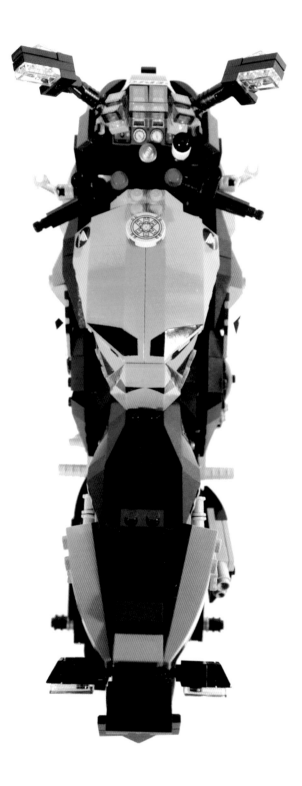

# KAWASAKI VULCAN 800

*Marat Andreev (2013)*

**+ Windshield**
Cut out from a plastic bottle

**+ Saddle bag**
Can be removed

**+ Brake disc**
Fixed to the front wheel

**+ V2 engine**
With moving pistons

**+ Folding stand**

## ABOUT THE MODEL

This model combines the smooth look of the real thing with the features and functions of an official Technic motorcycle: steering, full suspension, and a folding stand. Its piston engine is connected to the rear wheel via a chain, as well. Many features of the original bike are meticulously re-created, including the rear suspension with a single shock absorber concealed under the body panel and a front brake drum and brake disc complete with a brake cable. This model's windshield was painstakingly cut from a plastic bottle. Even the license plate is copied from an existing Kawasaki Vulcan!

## CHALLENGES

The builder's original plan was to use mostly custom-chromed pieces. However, the flexible hoses used to model the exhaust pipes proved too elastic to be coated in chrome. In the end, the model combines chrome and light grey, giving it a somewhat worn appearance.

### THE ORIGINAL

Introduced in 1995, the Kawasaki Vulcan combines the look of a traditional cruiser with some innovative technical solutions. These solutions include a mono-shock rear suspension, which later became common in sport bikes, and a liquid-cooled V-twin engine covered in cooling fins to maintain a traditional look. The 800 model was in production for 11 years and is still valued for its reliability and its classic sound.

**+ FUN FACT**

*The front brake disc is attached to the brake caliper and does not rotate with the wheel.*

*The hidden rear suspension, which is a feature also present in some Harley-Davidson models, is meant to make the motorcycle look like a hardtail chopper from the 1960s and 1970s.*

# SUPERCARS

# BUGATTI VEYRON 16.4 GRAND SPORT

*Sheepo (2010)*

## SPECIFICATIONS

| | |
|---|---|
| LENGTH | **22.8"** |
| WIDTH | **9.8"** |
| HEIGHT | **6.7"** |
| PIECES | **~3,200** |

*Even though the Bugatti Veyron sells for $1.6 million, various analysts and reporters have estimated that the vehicle costs about five times that to build. That means Volkswagen is losing more than $6 million on each Veyron it sells just to stay ahead of the competitors!*

## ABOUT THE MODEL

This model of the Bugatti Veyron supercar is built around an impressive remote-controlled 7+R sequential transmission. Designed for maximum functionality, the model also includes a W16 piston engine (which in this miniature version combines two V8 engines), an AWD drivetrain, working brakes in all wheels, a retractable rear wing that also tilts to act as an air brake when the brakes are engaged, and a motorized convertible roof.

The lightweight bodywork is made mostly with axles and elastic hoses plus a few panels. The doors and engine cover can be opened manually, while the hood is opened by a lever inside the cabin.

## CHALLENGES

For all its complex functionality, the Veyron is mostly about the transmission. It has the first sequential LEGO transmission with more than three speeds, so it's fairly complicated all by itself, and the mechanism that allows shifting with a single motor just adds to the complexity. Just making the transmission work with a heavy model was challenging, not to mention the challenge of fitting it inside the model.

Another difficulty was motorizing the innovative brake system, and then it had to be synchronized with the rear wing. All in all, this model took nine months of hard work.

## THE ORIGINAL

The Bugatti Veyron was built with a simple goal: to make sure that the Volkswagen Group, owner of the Bugatti brand, makes the fastest street-legal car in the world. But it took six years to create a production car fit for everyday use that was capable of speeds upward of 250 mph (400 km/h). The Veyron is fitted with a complex W16 engine, which is built like two intertwined V8s and delivers more than 1,000 hp, as well as 4 turbochargers and 10 radiators. Cruising at top speed for 12 minutes will dry fuel tanks and wear down a full set of tires, which are made just for the Veyron and can be replaced only in France.

Named "Car of the Decade" by **Top Gear**, the Bugatti Veyron has seen 4 regular and more than 20 special versions so far, and Volkswagen seems quite committed to react every time the competitors build a faster car.

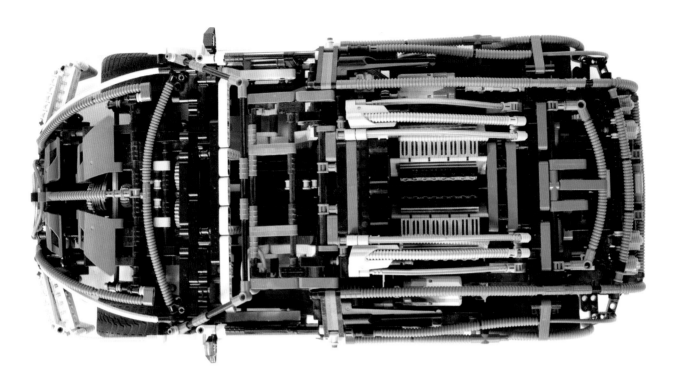

**+ FUN FACT**

The original Bugatti Veyron's roof
is removable, but only manually. It
can be taken off and stored inside
the car, and there is a fabric canopy
that can be opened quickly in case
of rain, but it's not recommended
for speeds over 80 mph (130 km/h).

# FERRARI 458 SPIDER

*brunojj1 (2012)*

**+** *Fully functioning convertible top*

## ABOUT THE MODEL

This compact model follows in the footsteps of several official LEGO Technic Ferrari sets like the Enzo set (#8653) and Fiorano set (#8145). While these provided inspiration, this Ferrari 458 has a few exciting surprises and improvements. The supercar comes with the features you'd expect, including a full independent suspension, remote-controlled propulsion, and a working steering wheel and lights. It doesn't have a transmission, but there is a much more unusual mechanism instead: the folding roof, which works seamlessly thanks to a PF M motor and a whole array of gear wheels.

## THE ORIGINAL

The 458 Italia (and its later Spider variant) was first announced as a radical new successor to the Ferrari F430 and featured a new design based on Ferrari's experience creating Formula 1 cars. The design includes a sophisticated suspension with advanced traction control systems and shock absorbers, whose damping power is controlled with electromagnets; brakes with a prefill function that can stop the car from 62 mph (100 km/h) to 0 in about 90 feet (27 meters); and a semi-automatic transmission that is also used by the Mercedes-Benz SLS AMG. All these solutions combined allowed the V8-powered 458 to lap the *Top Gear* test track just 0.1 seconds slower than the V12-powered Ferrari Enzo.

## SPECIFICATIONS

| | |
|---|---|
| LENGTH | **18.4"** |
| WIDTH | **9"** |
| HEIGHT | **4.7"** |
| PIECES | **2,077** |

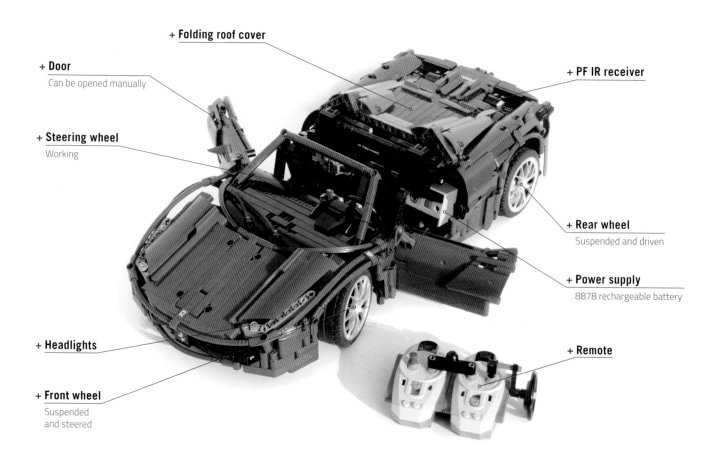

+ **Folding roof cover**

+ **Door**
Can be opened manually

+ **PF IR receiver**

+ **Steering wheel**
Working

+ **Rear wheel**
Suspended and driven

+ **Power supply**
8878 rechargeable battery

+ **Headlights**

+ **Remote**

+ **Front wheel**
Suspended
and steered

*For all its "street" looks, the real 458 is full of fancy technical inventions, including winglets in the front grille, which lower at high speeds to increase downforce. Former Ferrari Formula 1 driver Michael Schumacher was consulted for the interior design.*

# FORD MUSTANG SHELBY GT500

*Sheepo (2013)*

## SPECIFICATIONS

| | |
|---|---|
| LENGTH | **23.6"** |
| WIDTH | **9.7"** |
| HEIGHT | **6.9"** |
| PIECES | **4,006** |

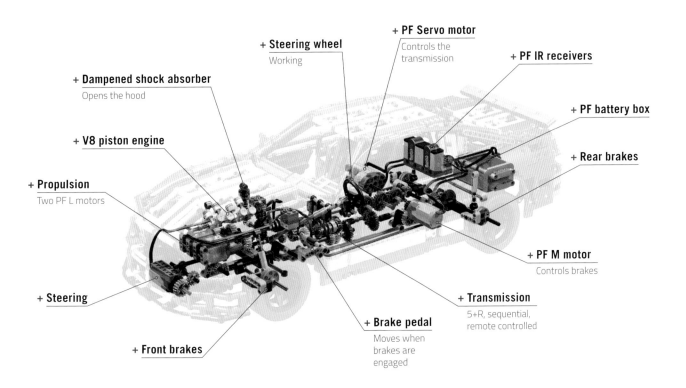

+ **Dampened shock absorber**
Opens the hood

+ **Steering wheel**
Working

+ **PF Servo motor**
Controls the transmission

+ **PF IR receivers**

+ **V8 piston engine**

+ **PF battery box**

+ **Propulsion**
Two PF L motors

+ **Rear brakes**

+ **Steering**

+ **PF M motor**
Controls brakes

+ **Front brakes**

+ **Brake pedal**
Moves when brakes are engaged

+ **Transmission**
5+R, sequential, remote controlled

## ABOUT THE MODEL

This 1:8 scale model of the Ford Mustang Shelby GT500 comes with a McPherson front suspension and live axle in the rear, as well as working disc brakes on all wheels, with a moving brake pedal inside the cabin. Front wheels have positive camber, with a caster and kingpin angle, and are steered with Ackermann geometry. The sequential 5+R transmission features an automated clutch, the hood opens to reveal a V8 piston engine, the trunk can be opened, and the doors even have locks!

The model consists of a beam-built unibody that can be taken off the modular chassis. This provides easy access for tinkering and adjustments and was helpful during the four months it took to develop the Mustang.

### CHALLENGES

The model was initially electrics-free. The decision to motorize meant a lot of work just to fit the electric components in it.

### THE ORIGINAL

The Shelby company has a long tradition of building souped-up versions of Ford Mustangs and producing them in limited numbers. Its work on the first-generation Mustang between 1965 and 1970 received worldwide recognition, and in 2007 the Shelby Company returned with a higher-performance variant of the renewed fifth-generation Mustang. The Shelby GT500 gets its name from a 500 hp engine and is also fitted with a tuned-up suspension, larger wheels, and a body kit.

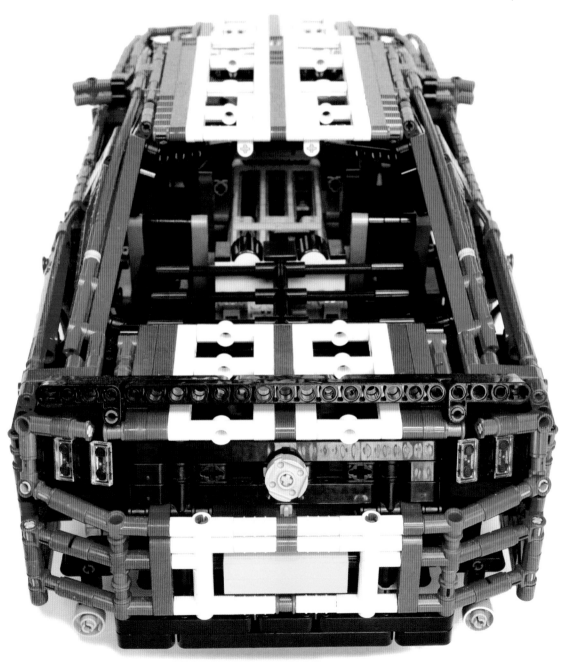

There is a special version of the GT500 called the American Shelby 1000 that is intended for track use only, with a 1,200 hp engine and a top speed estimated at 269 mph (430 km/h). To get one, you need to send your GT500 to the Shelby factory, where it will be stripped down and rebuilt from scratch.

# KOENIGSEGG CCX

*Jurgen Krooshoop (2012)*

## SPECIFICATIONS

| | |
|---|---|
| LENGTH | **17.4"** |
| WIDTH | **9"** |
| HEIGHT | **4.9"** |
| PIECES | **1,967** |

### ABOUT THE MODEL

This model of the exotic Swedish supercar is packed with features. Among them are a fully independent suspension, positive caster and camber angles on the front axle, a hand-of-God steering system with a working steering wheel and Ackermann geometry, a V8 piston engine, a six-speed transmission, and an opening hood and trunk. Most notably, the model reproduces the trademark Koeniseqq doors, which move outward and rotate at the same time.

### CHALLENGES

The doors! The unique opening motion required building a mechanism that could simultaneously move in two planes with a specific range, all the while fitting into a limited space.

### THE ORIGINAL

The first Koenigsegg to be street legal in the United States, the CCX, differs from the earlier two generations of this supercar by being compliant with most safety and environmental regulations worldwide. It's a splendid mix of state-of-the-art engineering, with an amazing 800 hp squeezed out of a V8 engine, and beautiful styling that made *Forbes* call it one of the world's 10 most beautiful cars. This incredibly fast car, with a top speed of 245 mph (395 km/h) is unique, with only 14 units built over four years.

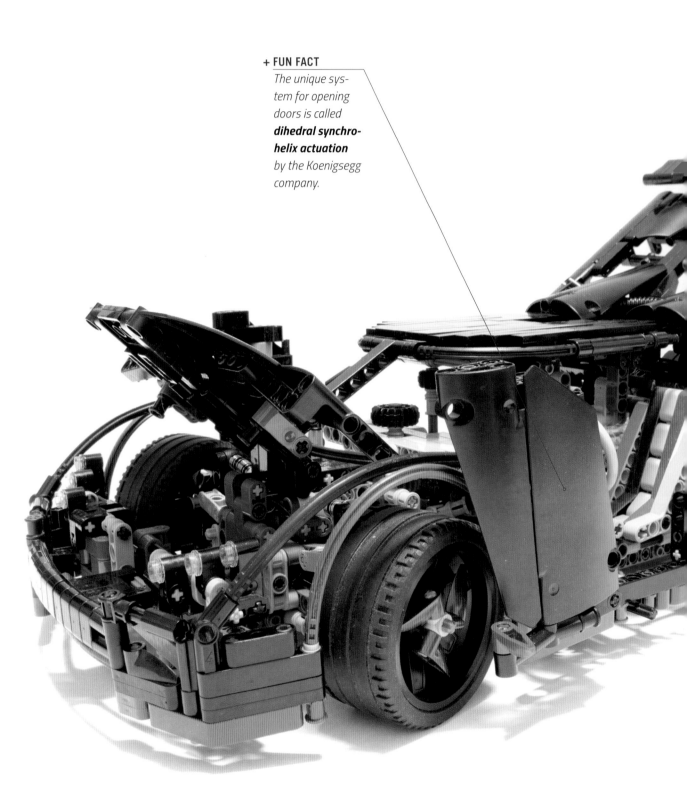

*There is a special variant of this car called CCXR, which runs on ethanol fuel and is hailed as the world's only "environmentally-friendly" supercar. The ethanol actually makes the engine more powerful; it develops up to 1,004 hp and increases the top speed up to 250 mph (402 km/h).*

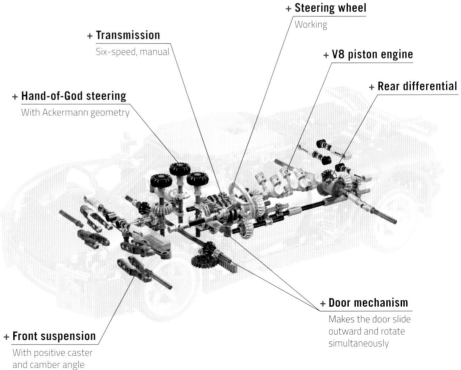

**+ Steering wheel**
Working

**+ Transmission**
Six-speed, manual

**+ V8 piston engine**

**+ Rear differential**

**+ Hand-of-God steering**
With Ackermann geometry

**+ Door mechanism**
Makes the door slide outward and rotate simultaneously

**+ Front suspension**
With positive caster and camber angle

# LAMBORGHINI AVENTADOR

*Francisco Hartley (2013)*

## ABOUT THE MODEL

This Aventador model has an intriguing development history. First, a fully functional chassis was created, and only after that was the body added. The model isn't motorized, which leaves it with plenty of space for mechanical functions, such as a full suspension, a V12 piston engine, an AWD system with a central differential, a 5+R manual transmission with a clutch, a working steering wheel, scissor doors opened with pneumatic shock absorbers, a working hood and trunk, and a retractable rear wing.

The carefully crafted body is mostly studless but also includes a number of bricks for accent details.

## CHALLENGES

The primary challenge was to make this model look authentic while following the "rules" of the modern LEGO Technic sets. These rules generally favor functions over looks, make every piece count, and allow you to hint at the iconic body shape, with gaps and holes, rather than build the body precisely.

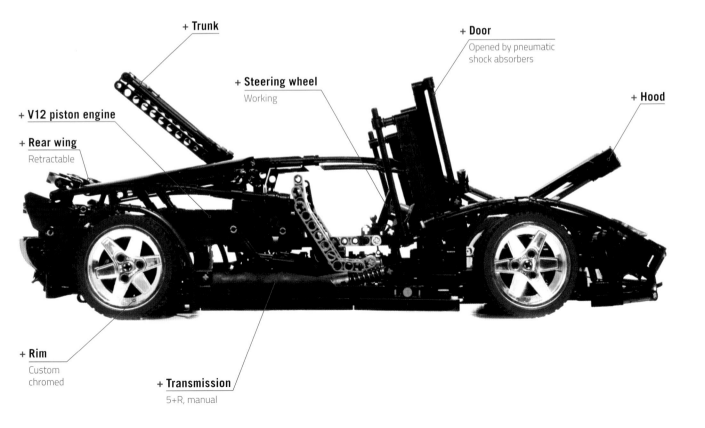

+ **Trunk**

+ **Steering wheel**
Working

+ **Door**
Opened by pneumatic shock absorbers

+ **Hood**

+ **V12 piston engine**

+ **Rear wing**
Retractable

+ **Rim**
Custom chromed

+ **Transmission**
5+R, manual

When tested on the **Top Gear** test track, the Aventador scored the then-third-fastest time, beating such cars as the Bugatti Veyron Super Sport, the Ferrari Enzo, and the Porsche 911 GT3.

### THE ORIGINAL

Even though Lamborghini's V12 flagship line was first introduced almost half a century ago, the Aventador is just the fifth car in the family. It was revealed 10 years after its predecessor, the Murciélago, and acclaimed for a number of advanced technological features, such as a carbon-fiber monocoque (shell) and a seven-speed ISR transmission.

Despite an incredible acceleration of 0 to 60 mph (97 km/h) in 2.9 seconds, the Aventador represents a new approach for Lamborghini V12 cars, with handling being given priority over top speed.

# LAMBORGHINI GALLARDO

*Crowkillers (2009)*

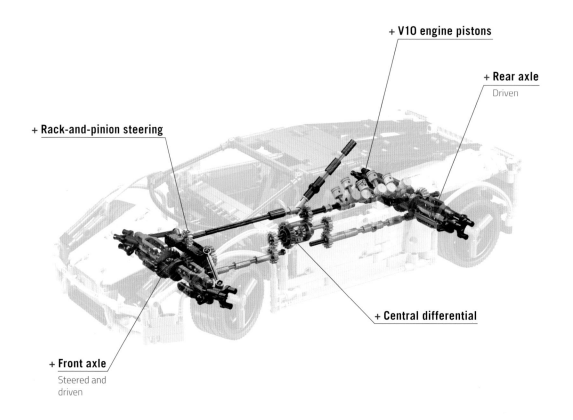

+ **V10 engine pistons**

+ **Rear axle**
Driven

+ **Rack-and-pinion steering**

+ **Central differential**

+ **Front axle**
Steered and driven

## ABOUT THE MODEL

This model, inspired by the official Technic sets, features an AWD drivetrain with a central differential and a mid-mounted V10 piston engine, hand-of-God steering, opening doors, and an opening engine bay cover, as well as custom-chromed rims. It succeeds in the difficult task of emulating the original car faithfully while using only the limited palette of pieces present in the modern LEGO Technic sets.

## THE ORIGINAL

Lamborghini Gallardo proved so popular that it was in production for 10 years, and it holds the record as Lamborghini's most produced car ever. Equipped with a 500 hp engine, a semi-automatic transmission that could operate fully automatically, and an exhaust flap to make the car quieter when driving in town, the Gallardo had more than 14,000 customers around the world.

Two Gallardos are being used by Italian traffic police, primarily for body organ transport but also for emergencies on Italian highways. Another two have been used temporarily by the Metropolitan Police of London, while one is currently employed by police in Panama.

# LAMBORGHINI MIURA JOTA

**Senator Chinchilla (2013)**

## SPECIFICATIONS

| | |
|---|---|
| LENGTH | **16.7"** |
| WIDTH | **7.9"** |
| HEIGHT | **4.25"** |
| PIECES | **~1,500** |

## ABOUT THE MODEL

This supercar-class model takes an unusual direction; it is built almost entirely from traditional LEGO bricks, including the chassis. Old-school as it may seem, the Lamborghini Miura Jota comes with a fully independent suspension, a steering system with a positive caster angle, a five-speed transmission, and a V12 piston engine. The doors, hood, and trunk can be opened; the headlights can be raised; and you can even adjust the seats by moving them forward or backward!

## CHALLENGES

This model had two essential challenges: fitting a transverse V12 piston engine and a five-speed transmission into a relatively small chassis and building a robust body with a massive hood and trunk that open in a rather odd way.

### THE ORIGINAL

Built in 1970, the Jota was a one-off version of the Lamborghini Miura. The original Miura was the first V12 Lamborghini flagship, followed by the legendary Countach, and it was known for starting the trend of two-seater, high-performance sports cars with a mid-engine layout. The Jota was its souped-up racing version, extensively modified to weigh a whopping 800 lb (360 kg) less and produce 90 hp more. It was sold to a private owner after being tested by the Lamborghini racing team, only to crash and burn completely the following year.

*The Miura was quite different from the cars Ferrucio Lamborghini wanted to build. The prototype shown at the 1966 Geneva car show made a great impression and ensured that Miura would see serial production.*

# MCLAREN MP4-12C

*Dikkie Klijn (2014)*

### ABOUT THE MODEL

This nonmotorized McLaren MP4-12C goes beyond re-creating just the exterior of the real car; it also houses a chassis replica. In a sense, it's a model-within-a-model, and it includes all the typical supercar functions, such as a full independent suspension, a V8 piston engine, a manual 3+R transmission, and a working steering wheel. The body combines two generations of Technic panels plus custom panels made with bricks and plates to get the right look, and its doors and engine cover can be opened.

The chassis, which includes scaled details such as the radiators, the exhaust system, and the engine, is modular as well. The entire model, which took 16 months to develop, can be easily separated into 25 modules but is quite robust when whole.

### THE ORIGINAL

Introduced in 2011, the McLaren MP4-12C, or simply 12C, was McLaren's first production car in 13 years and is the successor to its legendary F1 model. Making extensive use of Formula 1 technologies, the car is built around a one-piece carbon fiber monocoque that weighs just 175 lb (80 kg) and takes 4 hours to manufacture (it took 3,000 hours to produce the shell for the McLaren F1). Propelled by a 616 hp V8 engine, the 12C comes with an automatic seven-speed transmission including a manual preselect option, brake-assisted steering that reduces understeer by braking the inside wheel when cornering, and a set of brakes that can bring it from 125 mph (200 km/h) to zero in 5 seconds. It's one of the world's 10 fastest cars in terms of acceleration.

## SPECIFICATIONS

| | |
|---|---|
| LENGTH | **18.5"** |
| WIDTH | **8.5"** |
| HEIGHT | **5.1"** |
| PIECES | **~2,000** |

# MUSCLE CAR

*Crowkillers (2014)*

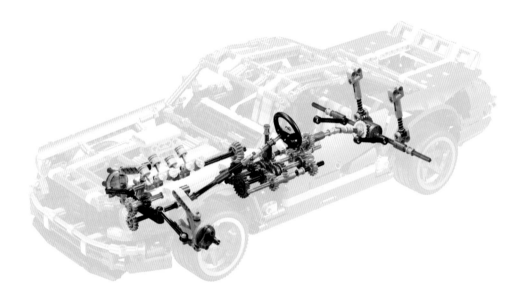

## SPECIFICATIONS

| | |
|---|---|
| LENGTH | **19.5"** |
| WIDTH | **8.3"** |
| HEIGHT | **5.8"** |
| PIECES | **1,887** |

## ABOUT THE MODEL

This love letter to the muscle cars of the 1960s features a four-speed transmission, an independent front suspension, a floating rear axle, a steering system with a working steering wheel, and a classic V8 piston engine.

Inspired by the Ford Mustang, Chevrolet Camaro, and Pontiac GTO, the car's body is carefully stylized, with a front chin spoiler, custom chromed rims, racing stripes, and a massive engine scoop. The doors and trunk can be opened, and the hood is based on Corvette hoods, which open forward and upward simultaneously.

### CHALLENGES

Styling this model had several challenges, such as making the transmission very flat to make the cabin's interior look realistic, shaping the roof, and making the hood open in just the right way with the air scoop sticking up through it.

# PAGANI ZONDA

*Sariel (2012)*

## SPECIFICATIONS

| | |
|---|---|
| LENGTH | **18.9"** |
| WIDTH | **9.4"** |
| HEIGHT | **5.5"** |
| PIECES | **~800** |

## ABOUT THE MODEL

Just like the original car, this model was built as a performance monster. To this end, it was built around four LEGO RC motors using a total of 12 AA batteries. All-wheel drive and an independent suspension are used to distribute the motors' combined power, with the chassis split into a front and rear unit, each with two RC motors coupled with an adder and driving a single axle. Steering is controlled by a PF M motor, while the ultralight body made of flexible axles helps keep the model's weight just less than 4.4 lb (2 kg). The model reaches a top speed of 9.6 mph (15.4 km/h) with fresh batteries, but they don't stay fresh for long.

## CHALLENGES

The primary challenge was lowering the weight and creating a sufficiently stress-resistant drivetrain. It was built with previously unused LEGO pieces, which proved durable enough.

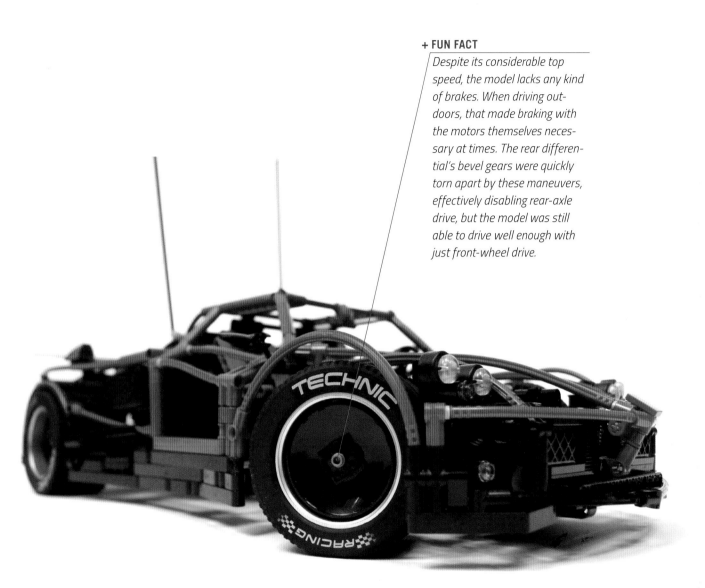

## + FUN FACT

*Despite its considerable top speed, the model lacks any kind of brakes. When driving outdoors, that made braking with the motors themselves necessary at times. The rear differential's bevel gears were quickly torn apart by these maneuvers, effectively disabling rear-axle drive, but the model was still able to drive well enough with just front-wheel drive.*

### THE ORIGINAL

Former Lamborghini employee Horacio Pagani set out to build a car of his own that would challenge supercars from companies such as Ferrari. He did exactly that with his first car, the hand-built Pagani Zonda C12. There have been many Zonda variants since then, but each of them amazed supercar fans with their unique blend of cutting-edge racing technology and vintage-feeling careful handcraftsmanship.

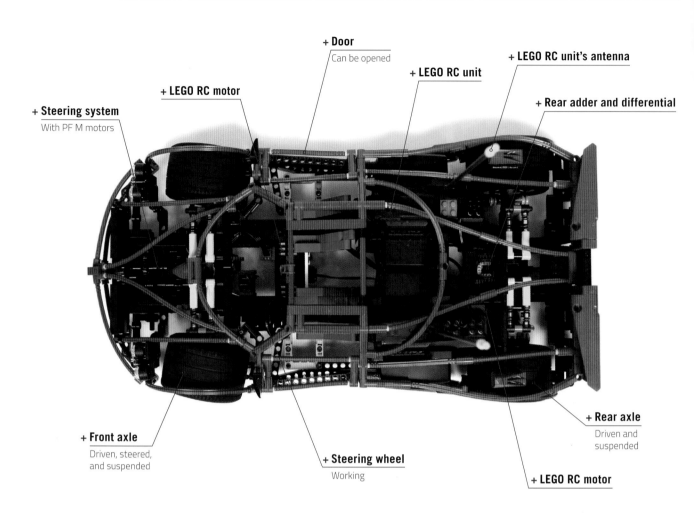

+ **Steering system**
With PF M motors

+ **LEGO RC motor**

+ **Door**
Can be opened

+ **LEGO RC unit**

+ **LEGO RC unit's antenna**

+ **Rear adder and differential**

+ **Front axle**
Driven, steered,
and suspended

+ **Steering wheel**
Working

+ **Rear axle**
Driven and
suspended

+ **LEGO RC motor**

Zonda is considered one of the world's most exotic cars, with just a few units of each variant produced. In fact, **serial** production is a bad word at Pagani because most Zondas are produced in 15 to 25 units, with some variants limited to 3 or 5 units, and there are a number of unique one-off editions.

# PORSCHE 911 (997) TURBO CABRIOLET PDK

*Sheepo (2011)*

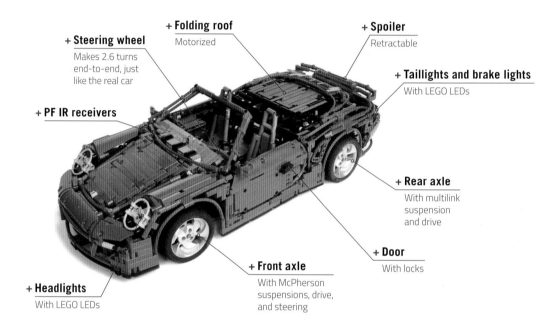

**+ Steering wheel**
Makes 2.6 turns
end-to-end, just
like the real car

**+ Folding roof**
Motorized

**+ Spoiler**
Retractable

**+ Taillights and brake lights**
With LEGO LEDs

**+ PF IR receivers**

**+ Rear axle**
With multilink
suspension
and drive

**+ Door**
With locks

**+ Front axle**
With McPherson
suspensions, drive,
and steering

**+ Headlights**
With LEGO LEDs

## SPECIFICATIONS

LENGTH **23.3"**

WIDTH **9.8"**

HEIGHT **6.9"**

PIECES **~4,000**

## ABOUT THE MODEL

This 1:7.5 model of the Porsche 911 was built
with accuracy in mind. Just like the real car, it
has a dual-clutch 7+R transmission, an H6 piston engine, AWD drive with a central differential, a McPherson front suspension, a multilink
rear suspension, a working steering wheel, and
disc brakes in all wheels, together with working
brake lights and a handbrake.

The body is richly detailed, including working
headlights and taillights, a hood operable by a
lever inside the cabin, and opening doors with
locks. Don't miss the retractable spoiler, the
motorized folding roof, and the fact that the
steering wheel does the same number of revo-
lutions when turning as it does in the real 911.
This model makes the real car look simple!

## CHALLENGES

The transmission was a challenge. It's nearly
8 inches (20 cm) long and made of more than
1,000 pieces, and it took eight months to get
it working. It's fully sequential and shifted by a
single motor, and it needs to be robust enough
to propel an 8 lb (3.6 kg) model. Building it was
just part of the challenge; building a scaled
model around it wasn't any easier.

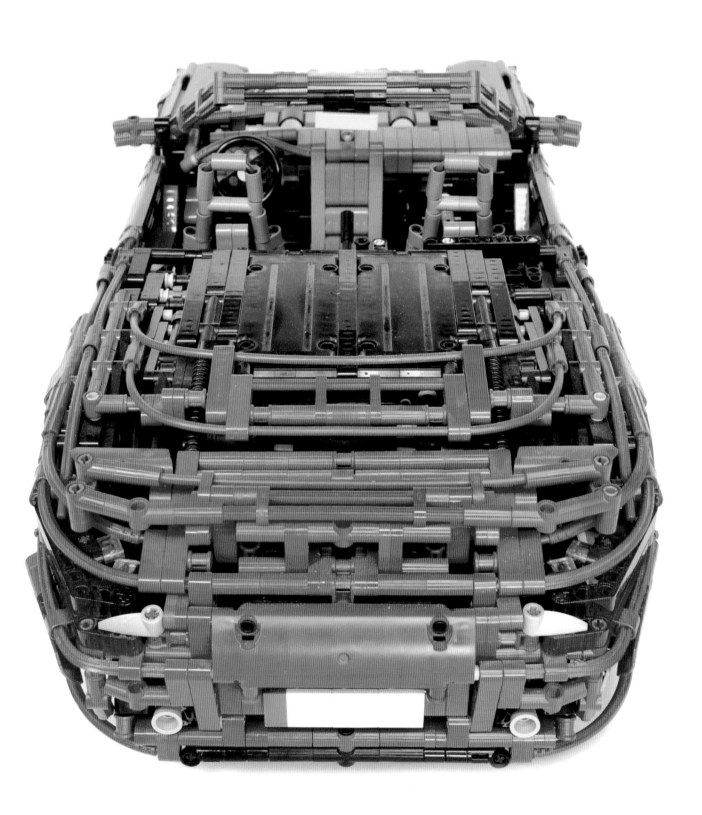

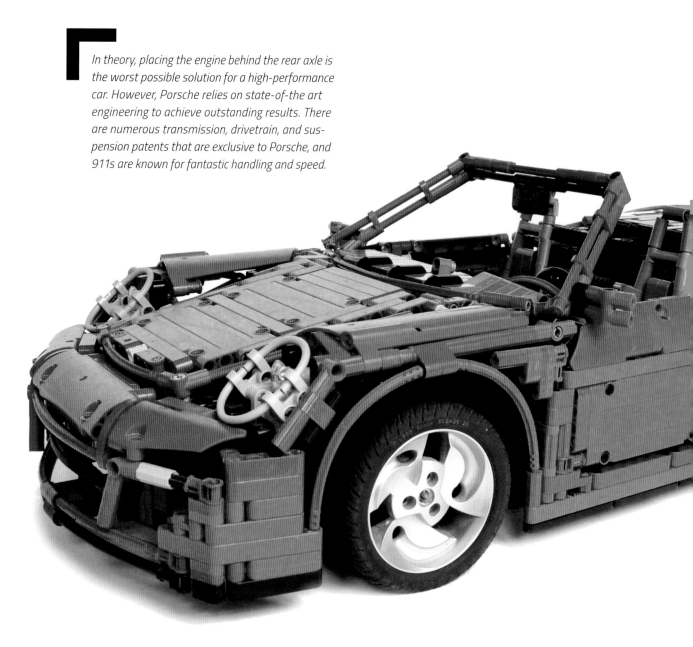

*In theory, placing the engine behind the rear axle is the worst possible solution for a high-performance car. However, Porsche relies on state-of-the art engineering to achieve outstanding results. There are numerous transmission, drivetrain, and suspension patents that are exclusive to Porsche, and 911s are known for fantastic handling and speed.*

### THE ORIGINAL

The 911 is the Porsche flagship sports car, the first version of which was designed in 1959 and debuted in 1963. It was so successful that it saw little change in 26 years on the market. It was followed by six other generations, with the basic concept holding throughout: 2+2 seats and a rear-mounted H6 engine. The latest generation's engine has more than 500 hp, compared to the first 911s' 130 hp.

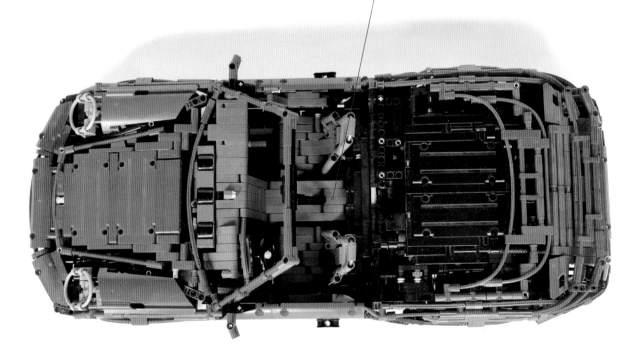

+ **FUN FACT**

*This model includes more than three times as many pieces as the LEGO Technic supercar released the same year: set #8070. Clearly, the LEGO Group has some catching up to do.*

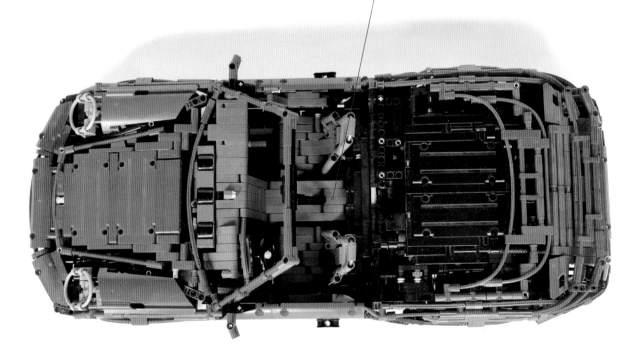

# VAMPIRE GT

*Crowkillers (2012)*

## SPECIFICATIONS

| | |
|---|---|
| LENGTH | **17.4"** |
| WIDTH | **8"** |
| HEIGHT | **4.7"** |
| PIECES | **1,925** |

**+ FUN FACT**

*The original black model and a later white variant were both sold to raise money for charity.*

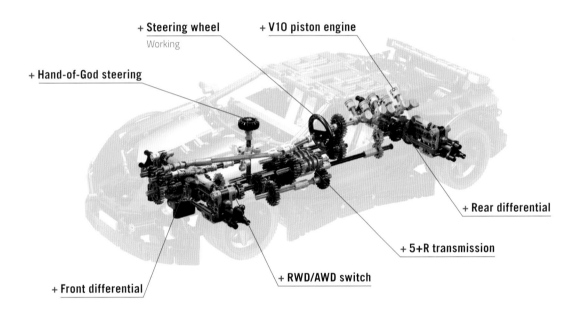

+ Steering wheel
Working

+ V10 piston engine

+ Hand-of-God steering

+ Rear differential

+ 5+R transmission

+ RWD/AWD switch

+ Front differential

## ABOUT THE MODEL

This model of a fictional supercar comes equipped with a fully independent suspension, a 5+R transmission, a central differential, a drivetrain that can be switched between RWD and AWD modes, a V10 piston engine, gull-wing doors, and hand-of-God steering with a working steering wheel. The Vampire has a panel-built body full of interesting shapes, giving it a menacing look. The hood and trunk open too, and the custom-chromed red rims make sure you have the best-looking car in your neighborhood.

## CHALLENGES

The biggest challenge was the transmission, with a shift pattern taken from cars such as the Ford Mustang or Subaru Impreza, where the reverse gear comes after the fifth gear rather than before the first one. This required making sure the first and reverse gears have similar ratios, even though they're located on opposite sides of the transmission.

# TRACKED VEHICLES

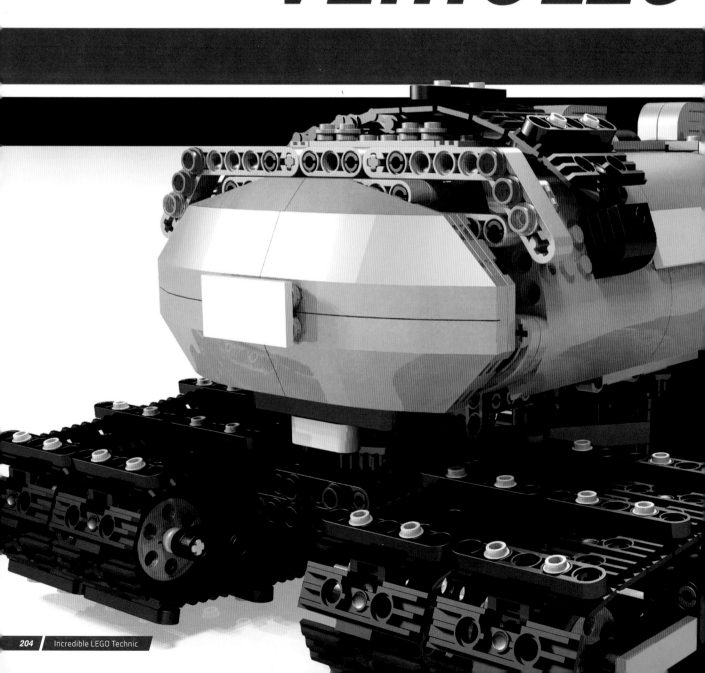

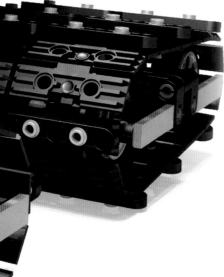

# BRIDGELAYER

*Mahj (2011)*

## ABOUT THE MODEL

This bridgelayer was built with the goal of deploying a bridge using a remote control only, then having the model cross it, and finally recovering the span. Two motors are used to power its tracks, while another two lay the bridge down and power it so it can unfold. The carrier is then detached from the bridge by simply driving backward, making it free to drive across the bridge.

Once on the other side, it turns around and lowers a claw that connects it to the bridge. The deployment process is then reversed, placing the folded bridge on top of the carrier, and the model can continue driving.

## CHALLENGES

The primary challenge was the bridge itself, which has to unfold and fold to be deployed from one end and picked back from the other. The folding is performed by two linear actuators driven by a motor inside the carrier. The ends of the bridge are symmetric, and the bridgelayer's claw can slide into either one, creating a connection strong enough to pick up the bridge while also driving its folding mechanism.

## SPECIFICATIONS

| | |
|---|---|
| LENGTH | **14.5"** |
| WIDTH | **5.7"** |
| HEIGHT | **5.7"** |

### THE ORIGINAL

Bridgelayers are used exclusively by military forces to help armored units cross rivers, ditches, and similar obstacles quickly, even under combat conditions. They are usually based on the chassis of regular tanks, fitted with armor of their own but mostly unarmed. They were introduced in World War I, and some modern models can span more than 300 feet (100 meters) and handle vehicles that weigh more than 60 tons.

*When a bridge is raised vertically during the deployment process, it can be seen from afar. Some bridge-layers designed for stealthy operations deploy the bridge by sliding it off the carrier horizontally.*

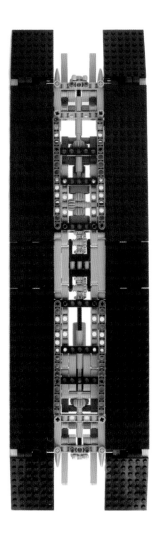

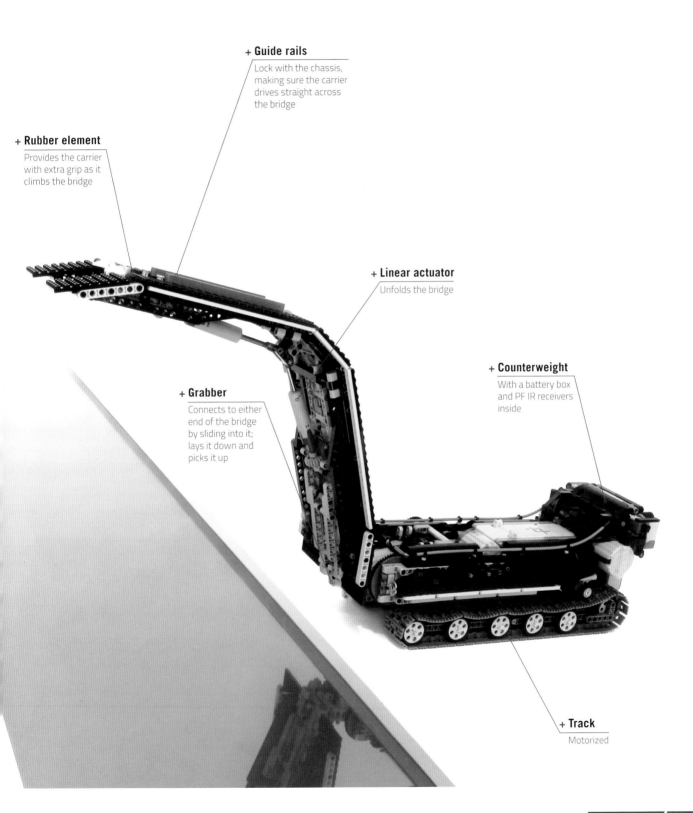

**+ Guide rails**
Lock with the chassis, making sure the carrier drives straight across the bridge

**+ Rubber element**
Provides the carrier with extra grip as it climbs the bridge

**+ Linear actuator**
Unfolds the bridge

**+ Counterweight**
With a battery box and PF IR receivers inside

**+ Grabber**
Connects to either end of the bridge by sliding into it; lays it down and picks it up

**+ Track**
Motorized

# K2 BLACK PANTHER

*Sariel (2013)*

**SPECIFICATIONS**

| | |
|---|---|
| LENGTH | **18.9"** |
| WIDTH | **8.1"** |
| HEIGHT | **6.3"** |
| PIECES | **~2,500** |

## ABOUT THE MODEL

This model of the South Korean main battle tank—called K2 Black Panther—sports an impressive suspension system. Each of the 12 road wheels in the treads is suspended on a torsion bar that can be rotated remotely at any moment. This means the suspension of each individual track can be raised or lowered at will, even while traversing obstacles. Thanks to a powerful propulsion system, the model is also agile and equipped with a turret whose front opens to reveal a concealed LEGO spring cannon that can be fired remotely. Turret rotation and the main gun's elevation are, of course, also controlled remotely.

## CHALLENGES

The model is extremely compact, with the hull housing a huge adjustable suspension system, two PF XL motors, three PF IR receivers, a PF battery, and a single PF M motor with a turret rotation mechanism. The turret houses just one PF M motor, equipped with a distribution gearbox that allows it to elevate the main gun or to fire the spring cannon. It was challenging to fit all these elements inside the model while keeping it as flat as the real tank. In the end, I think I actually made it a bit too flat!

## THE ORIGINAL

The K2 is a cutting-edge design of modern military technology. The tank's features include an autoloader capable of firing a new round every 3 seconds; a fully active suspension system that allows the whole tank to "sit down" or tilt in any direction; state-of-the art detection, jamming, and countermeasure electronic systems; ammunition with a tungsten core; and the ability to engage targets beyond the line of sight by firing a special type of fire-and-forget round at a mortar-like trajectory.

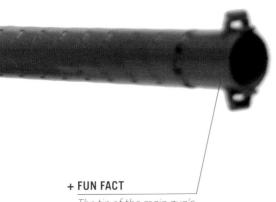

### + FUN FACT

*The tip of the main gun's barrel is made of a trash can for LEGO minifigures.*

**+** *Fully functioning  suspension elements*

The K2 is the world's only tank that can be fitted with a 140 mm caliber gun with only minimal modifications. Its development program includes tests with electrothermal-chemical guns that may replace conventionally propelled rounds in the future.

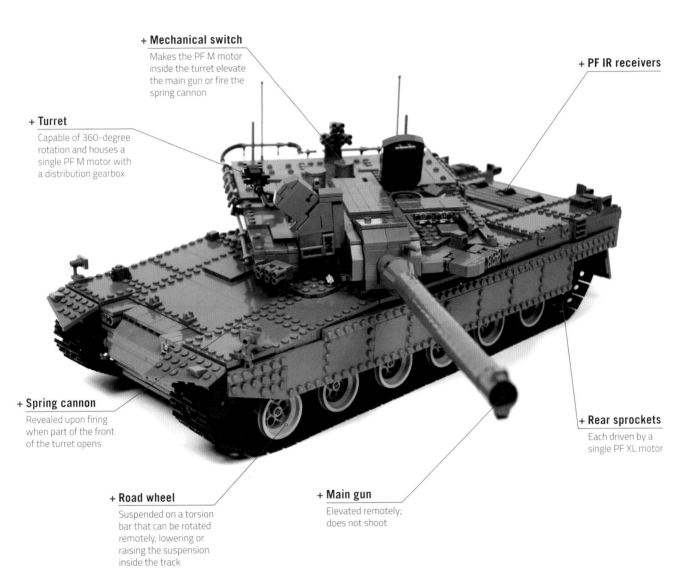

**+ Mechanical switch**

Makes the PF M motor
inside the turret elevate
the main gun or fire the
spring cannon

**+ PF IR receivers**

**+ Turret**

Capable of 360-degree
rotation and houses a
single PF M motor with
a distribution gearbox

**+ Spring cannon**

Revealed upon firing
when part of the front
of the turret opens

**+ Rear sprockets**

Each driven by a
single PF XL motor

**+ Road wheel**

Suspended on a torsion
bar that can be rotated
remotely, lowering or
raising the suspension
inside the track

**+ Main gun**

Elevated remotely;
does not shoot

# LAND RAIDER

*Jerac (2012)*

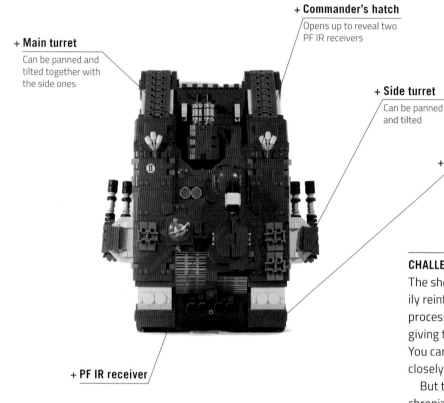

**+ Main turret**
Can be panned and tilted together with the side ones

**+ Commander's hatch**
Opens up to reveal two PF IR receivers

**+ Side turret**
Can be panned and tilted

**+ Rear sprocket**
Driven by two coupled PF XL motors

**+ PF IR receiver**

## ABOUT THE MODEL

This minifig-scale model of the Land Raider tank from **Warhammer 40,000** looks real enough to be mistaken for an actual game model—that is, until it starts moving. Propelled by four PF XL motors and weighing more than 7.5 lb (3.5 kg), it's massive enough to push small pieces of furniture around.

The main turret can move sideways and up and down, and so can the two side turrets, which are synchronized with the main turret. The model, which took eight months to develop, is also fitted with working headlights and covered by a body that includes every detail of the original vehicle.

## CHALLENGES

The sheer weight of the model required a heavily reinforced drivetrain. Through the design process, the model broke many gear wheels, giving the model its nickname, Geargrinder. You can see the broken gears count if you look closely under its name on the nose.

But the true challenge was the turret synchronization mechanism. All three turrets are operated by just two PF M motors, and each turret has a limiter to stop it when it reaches the extreme position, while the other turrets go on moving. This required a complex mechanism that involved rubber bands and plenty of gear wheels.

## THE ORIGINAL

One of the largest tracked vehicles in **Warhammer 40,000**, the Land Raider acts as a tank and troop transport at the same time. Heavily armored and impressively armed, it's more than capable of delivering troops to any part of the battlefield and providing them with devastating fire support. It even comes with an AI driver to keep it operational with no crew!

| | |
|---|---|
| LENGTH | **14.8"** |
| WIDTH | **8.5"** |
| HEIGHT | **5.35"** |
| PIECES | **3,100** |

# PRINOTH LEITWOLF

*Designer-Han (2007)*

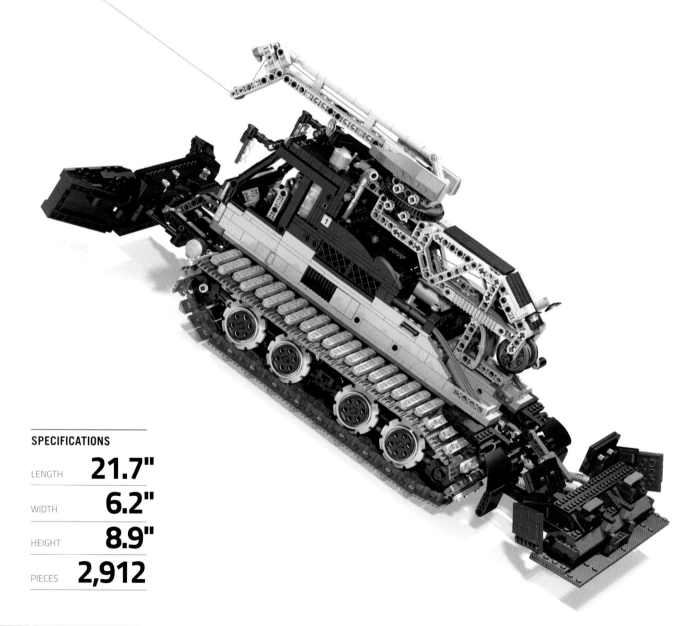

## SPECIFICATIONS

| | |
|---|---|
| LENGTH | **21.7"** |
| WIDTH | **6.2"** |
| HEIGHT | **8.9"** |
| PIECES | **2,912** |

### ABOUT THE MODEL

This snow groomer model comes with two PF XL motors driving the rear sprockets, a pneumatic system that allows the adjustment of the ground clearance of the front sprockets, a V6 piston engine, a front blade that can be lowered, and a rear tiller equipped with rotating cutter rollers. There is also a manually operated winch with a boom in the back. The chassis is fitted with two sets of tracks on each side; the half-stud beams added to the tracks give them a better grip in snow.

### CHALLENGES

The primary challenges were placing all the electric components safely in the chassis so they would not come in contact with the snow and removing any slack from the tracks, which was done by suspending the front sprockets on shock absorbers. It was also difficult to make the chassis robust enough while leaving enough space inside for all the functions.

### THE ORIGINAL

Leitwolf is one of the premium snow groomer models from the Italian Prinoth company and is used to prepare ski trails. Driven by a 435 hp engine, it is specifically designed for extreme weather conditions, and it comes with a winch that can function as an anchor and has a reach of 0.5 miles (850 meters).

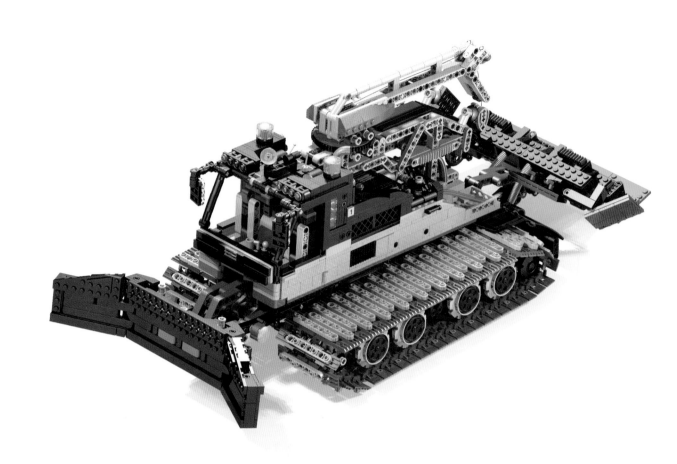

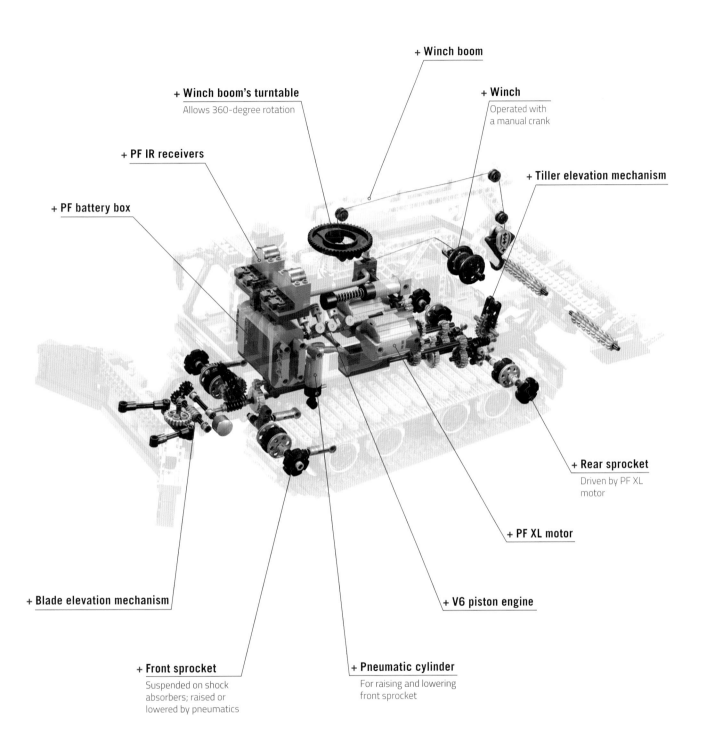

+ **Winch boom**

+ **Winch boom's turntable**
Allows 360-degree rotation

+ **Winch**
Operated with
a manual crank

+ **PF IR receivers**

+ **Tiller elevation mechanism**

+ **PF battery box**

+ **Rear sprocket**
Driven by PF XL
motor

+ **PF XL motor**

+ **V6 piston engine**

+ **Blade elevation mechanism**

+ **Front sprocket**
Suspended on shock
absorbers; raised or
lowered by pneumatics

+ **Pneumatic cylinder**
For raising and lowering
front sprocket

# STARCRAFT SIEGE TANK

*drakmin (2012)*

## SPECIFICATIONS

| | |
|---|---|
| LENGTH | **24.5"** |
| WIDTH | **18"** |
| HEIGHT | **9"** |
| PIECES | **~5,000** |

### ABOUT THE MODEL

This massive model uses 11 motors and an intricate pneumatic system to re-create functions of the iconic siege tank from the **StarCraft** game universe. The functions include six motorized tracks, a rotating turret, and an extending main gun that fires two spring cannons. But most impressive is the faithful re-creation of the transformation between "tank" and "siege" modes. To complete the transformation, two complex outriggers move out from inside the middle tracks, while the whole front and rear tracks' assemblies extend sideways and downward to lift the model.

### CHALLENGES

Making a model that weighs approximately 11 lb (5 kg) lift itself on a complex chassis with multiple moving parts took an entire year to work out. The hardest part was making the middle tracks motorized while retaining the ability to pull them inside the chassis and extend the outriggers right through their housings.

### THE ORIGINAL

One of the most fearsome units in the Terran faction, the siege tank can operate like a regular tank or like a long-range immobile artillery piece with massive splash damage when in "siege mode."

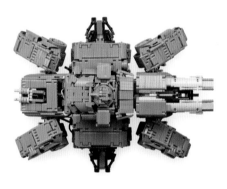

The siege tank in **StarCraft II** has two looks: The one shown in the game menu is stream-lined and curvy, while the in-game model is bulkier. The model aims to combine the best of both looks.

# STILZKIN INDRIK

*Mahj (2010)*

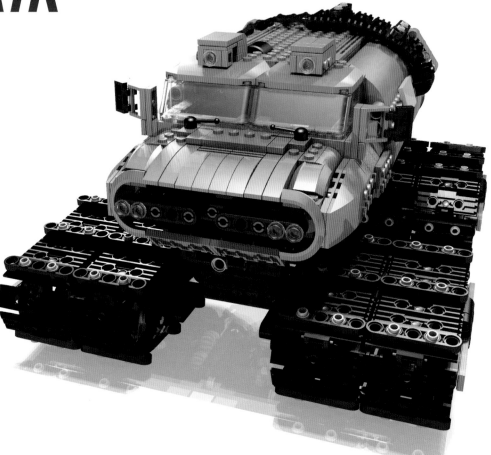

### ABOUT THE MODEL

This model of a fictional arctic exploration vehicle features four tracks driven by two independent motors and steered by slewing, as well as powerful lights for nighttime operations. It's not fast, but it climbs through deep snow and pulls two fully loaded cargo sledges without any difficulty.

The chassis is robust, stable, and tall enough to keep all sensitive electric elements far from snow, and the body is lightweight and fully closed to keep the innards dry and cozy.

### THE ORIGINAL

Inspiration for the model comes from the Tucker Sno-Cat vehicles used to explore the Arctic and Antarctic regions in the 1950s. The orange body adds authenticity—arctic vehicles are usually painted orange to stand out against their white surroundings.

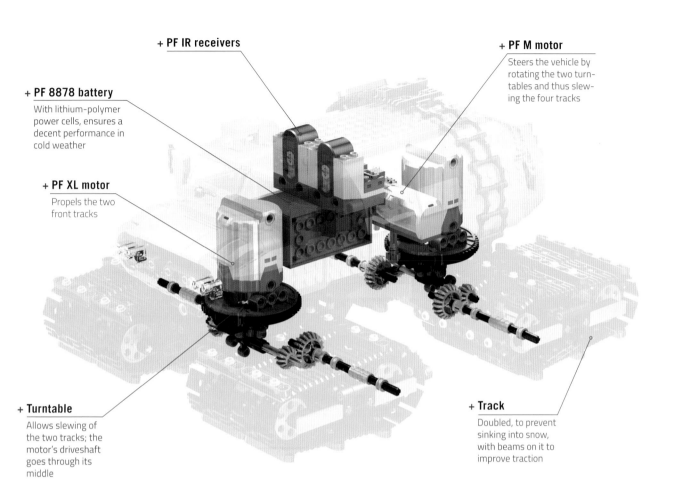

**+ PF IR receivers**

**+ PF 8878 battery**
With lithium-polymer power cells, ensures a decent performance in cold weather

**+ PF M motor**
Steers the vehicle by rotating the two turntables and thus slewing the four tracks

**+ PF XL motor**
Propels the two front tracks

**+ Turntable**
Allows slewing of the two tracks; the motor's driveshaft goes through its middle

**+ Track**
Doubled, to prevent sinking into snow, with beams on it to improve traction

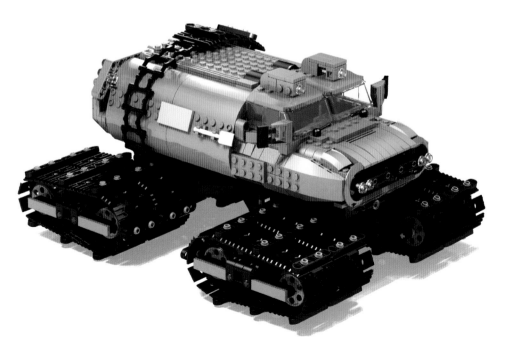

### SPECIFICATIONS

| | |
|---|---|
| LENGTH | **11.3"** |
| WIDTH | **9.1"** |
| HEIGHT | **6.3"** |
| PIECES | **1,250** |

# WHITE TIGER T1H1

*GYUTA (2010)*

## SPECIFICATIONS

LENGTH **22.8"**

WIDTH **10.2"**

HEIGHT **7"**

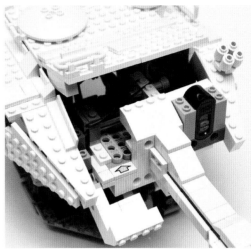

### ABOUT THE MODEL

While most LEGO tank models are built with a focus on driving performance, suspension, and aesthetics, this one comes armed up to the teeth. This tank's cannon is not only functional but even has an autoloader that holds four shells. The remotely operated turret contains its own power source, and the firing mechanism within is similar to an airsoft gun.

The model is propelled by two PF XL motors and steered by a third one through a mechanism that combines an adder and a subtractor for easy control. Just as in a real tank, the propulsion system can be easily removed along with the partial drivetrain. All road wheels are suspended independently, and the side skirts can be taken off.

### CHALLENGES

The firing and loading mechanisms were obviously challenging to develop, but so was the propulsion system. The two separate power supplies make the model quite heavy, and it was difficult to sufficiently reinforce the drivetrain while keeping it compact and removable.

### THE ORIGINAL

This model is loosely inspired by the South Korean K2 Black Panther (see page 210).

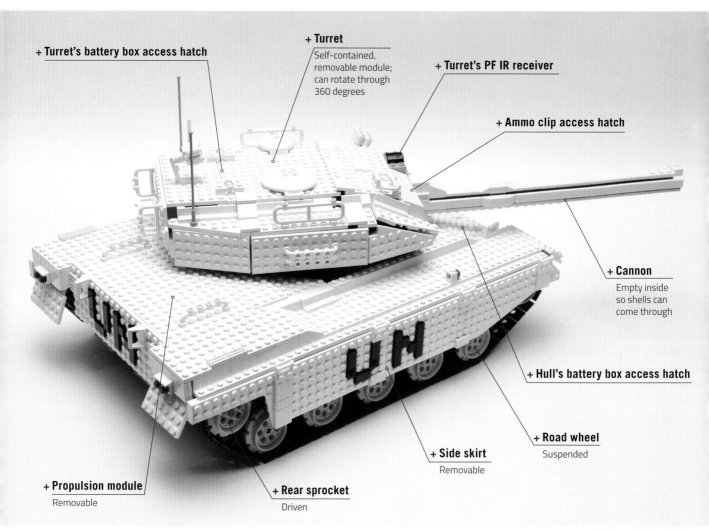

+ **Turret's battery box access hatch**

+ **Turret**
Self-contained, removable module; can rotate through 360 degrees

+ **Turret's PF IR receiver**

+ **Ammo clip access hatch**

+ **Cannon**
Empty inside so shells can come through

+ **Hull's battery box access hatch**

+ **Road wheel**
Suspended

+ **Side skirt**
Removable

+ **Rear sprocket**
Driven

+ **Propulsion module**
Removable

*All vehicles used by the United Nations are completely white. This is a symbolic choice, as well as a practical one, because it makes sense for peacekeeping forces to make their presence known.*

# TRUCKS

# AMERICAN TRUCK

*2LegoOrNot2Lego (2013)*

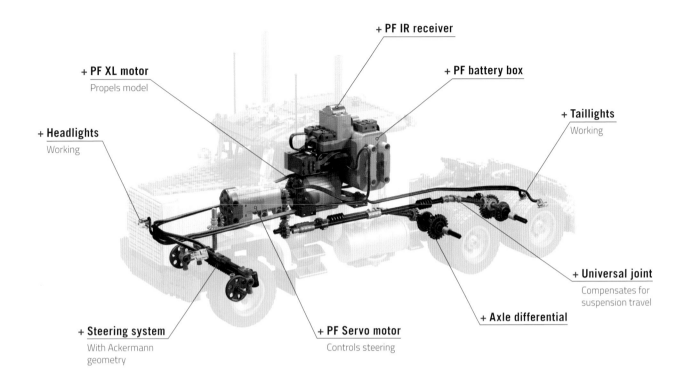

+ PF IR receiver

+ PF XL motor
Propels model

+ PF battery box

+ Taillights
Working

+ Headlights
Working

+ Universal joint
Compensates for
suspension travel

+ Steering system
With Ackermann
geometry

+ PF Servo motor
Controls steering

+ Axle differential

## SPECIFICATIONS

| | |
|---|---|
| LENGTH | **18.9"** |
| WIDTH | **6.6"** |
| HEIGHT | **9.1"** |
| PIECES | **1,783** |

## ABOUT THE MODEL

This model combines features of Kenworth, Peterbilt, Mack, Autocar, Western Star, and Freightliner trucks to represent an iconic American big rig. This truck is both functional and attractive—it's fitted with a full suspension and Ackermann steering, it's motorized and remote-controlled, and it has a model of the Detroit DD15 diesel engine under the hood. The sophisticated four-color livery adds to the effect and is the builder's trademark. The fifth wheel locks automatically when you attach a trailer.

## CHALLENGES

Standard LEGO shock absorbers proved too large to create a working suspension without affecting the model's overall design. They were replaced with rubber belts, and it took many trials to create a compact and effective suspension system.

The term **fifth wheel** (the device that connects
the trailer to the truck) originates from the time of
horse-drawn carriages and wagons. To enable a
carriage with a wagon to make a turn, the connection between them must pivot on a horizontal plane.
Early carriages simply had an extra wheel placed flat
on the back of their frames as a connecting point,
and since the carriages were typically four-wheeled,
the term **fifth wheel** was coined.

# DUMP TRUCK 10×4

*Designer-Han (2009)*

### ABOUT THE MODEL

The model is based on a Dutch Ginaf X5450 five-axle dump truck. The three front axles are steered, and the two rear axles are connected to a V8 piston engine by a realistic drivetrain. A manually operated pneumatic system tips the dump bed and lifts the third axle. The chassis is robust and offers impressive load capacity: The truck rolls and steers, even with the bed full of LEGO pieces.

### CHALLENGES

Just like in the real truck, lifting the third axle disconnects it from the steering system and at the same time locks it in a "straightforward" position. Re-creating this action was the most difficult aspect of this build.

### THE ORIGINAL

Ginaf is a Dutch truck manufacturer that works with DAF, which manufactures the engines, drivetrains, and cabins. The X5450 truck was released in 2003, with four of its five axles driven by a powerful six-cylinder inline engine. The last rear axle can also steer, resulting in superb steering capabilities.

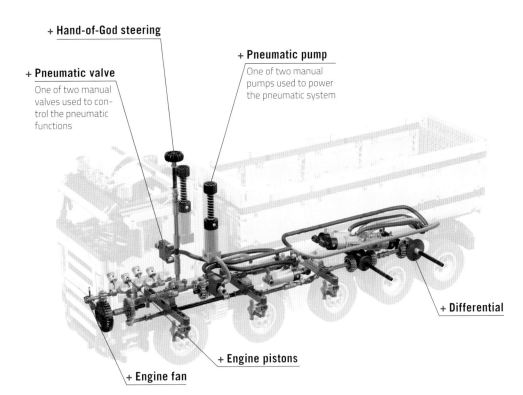

**+ Hand-of-God steering**

**+ Pneumatic valve**
One of two manual valves used to control the pneumatic functions

**+ Pneumatic pump**
One of two manual pumps used to power the pneumatic system

**+ Differential**

**+ Engine pistons**

**+ Engine fan**

## SPECIFICATIONS

| | |
|---|---|
| LENGTH | **21"** |
| WIDTH | **7.1"** |
| HEIGHT | **8.7"** |
| PIECES | **2,454** |

**+ FUN FACT**
*The complexity of the pneumatic system and the amount of pressurized air present inside it at all times make the bed tip down gradually, just like in the real truck.*

# KAMAZ DAKAR RALLY TRUCK

*Marat Andreev (2014)*

## ABOUT THE MODEL

This model, built as a tribute to the "Kamaz Master" Dakar rally team, combines authenticity and performance while also demonstrating the builders' skill with custom stickers. Functions include a 4×4 drive, steering, and a complex live axle suspension stabilized with Watt linkages. Designed for speed rather than climbing, the model is propelled by two PF M motors carefully balanced and kept as light as possible. The suspension's hardness can be adjusted, and its travel is truly impressive.

## CHALLENGES

The livery, which was carefully re-created from the original truck, proved to be challenging not only because the documentation for the original Kamaz Master livery was lacking but also because it had to include two tones of blue LEGO pieces to match the model's exterior.

## THE ORIGINAL

Kamaz is the biggest truck manufacturer in Russia. Its racing team, founded in 1988, quickly became one of the leaders in the famous Paris Dakar rally, a grueling endurance race.

The racing truck's design has a mid-engine V8 producing 850 hp, a reinforced chassis, and military-grade suspension components. The truck has survived 39-foot (12-meter) long jumps and is capable of reaching top speeds up to 100 mph (165 km/h) off-road. The Kamaz Master team currently holds 12 Dakar race victories.

# KENWORTH 953 OILFIELD TRUCK

*2LegoOrNot2Lego (2011)*

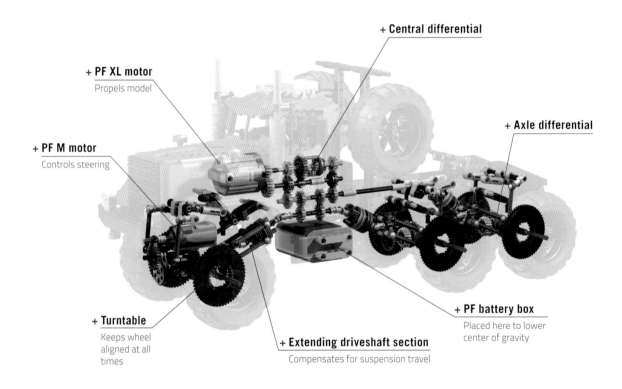

**+ Central differential**

**+ PF XL motor**
Propels model

**+ Axle differential**

**+ PF M motor**
Controls steering

**+ Turntable**
Keeps wheel aligned at all times

**+ Extending driveshaft section**
Compensates for suspension travel

**+ PF battery box**
Placed here to lower center of gravity

## SPECIFICATIONS

| | |
|---|---|
| LENGTH | **20.3"** |
| WIDTH | **7.8"** |
| HEIGHT | **9.4"** |
| PIECES | **1,654** |

## ABOUT THE MODEL

This model of a Kenworth 953 6×6 semi-truck features a complex suspension system with three floating axles and a robust drivetrain with a central differential and extending driveshaft sections. All wheels are driven and mounted on turntables, which keep them aligned and which act as bearings. The fifth wheel locks automatically when a trailer is attached.

## CHALLENGES

The model faced typical challenges of an off-road truck: finding a compromise between high ground clearance and good stability, making sure the suspension has long travel and can handle the model's weight, and reinforcing the drivetrain.

### THE ORIGINAL

The Kenworth 953 Oilfield Truck was designed to handle heavy loads in difficult terrain and under extreme weather conditions. Standing nearly 13 feet (4 meters) tall, this monstrous vehicle runs on six wheels, each taller than a person and weighing 2,000 lb (900 kg) together with the tire. Under the enormous hood is a 457 hp 12V diesel engine. Rough, tough, and fitted with enormous air filters, an air conditioning unit, and fuel tanks with a capacity of 550 gallons, this truck is frequently used in the deserts of Saudi Arabia. It is, as Kenworth's motto puts it, "built for the job."

# KENWORTH W900L DUMP TRUCK

*M_Longer (2012)*

## ABOUT THE MODEL

Kenworth trucks are popular among Technic builders, and this model proves you can't have a proper Kenworth truck without an abundance of custom chromed pieces. Driven and steered remotely, it also comes with a motorized dump mechanism, a second axle that can be lifted, and working headlights. To make room for the cabin interior, the battery and PF IR receivers are hidden under the hood.

## CHALLENGES

The weight of the detailed body and cargo bay puts serious pressure on the chassis. That made it difficult to build a frame that would be rigid enough for a model this long and to develop a dump mechanism that would be strong enough to work and yet compact enough to fit under the cargo bay.

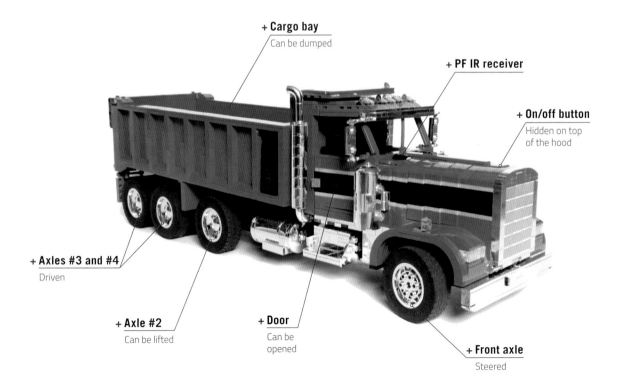

+ **Cargo bay**
Can be dumped

+ **PF IR receiver**

+ **On/off button**
Hidden on top of the hood

+ **Axles #3 and #4**
Driven

+ **Axle #2**
Can be lifted

+ **Door**
Can be opened

+ **Front axle**
Steered

## SPECIFICATIONS

| | |
|---|---|
| LENGTH | **21.6"** |
| WIDTH | **5.9"** |
| HEIGHT | **7.1"** |

LONGER

### THE ORIGINAL

The W900 model (the **W** stands for Worthington) is a classic Kenworth truck designed for long-distance hauling. With a choice of engines that develop up to 625 hp and a variety of customization options for the body, the W900 makes sure the driver feels at home and the goods are delivered on time.

# KZKT-7428 RUSICH

*Sariel (2013)*

## SPECIFICATIONS

LENGTH **69"**

WIDTH **8.2"**

HEIGHT **8.5"**

PIECES **~6,000**

## ABOUT THE MODEL

The KZKT-7428 Rusich model was built specifically to tow my K2 Black Panther tank model (see page 200). Almost 6 feet (1.8 meters) long with the trailer attached, this model features an 8×8 drive, an independent suspension, and a remotely locked fifth wheel. Rounding things out are the trailer's rear ramp, which can move up and down, and its outriggers. All of this is controlled via a Bluetooth link, using Mindstorms NXT brick-reading commands from a gaming pad connected to a computer. This allows both the tank carrier and the IR-controlled tank model to operate at the same time without interference.

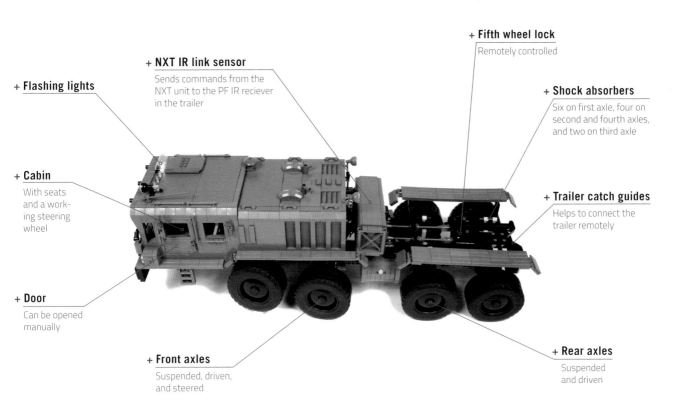

**+ Flashing lights**

**+ NXT IR link sensor**
Sends commands from the NXT unit to the PF IR reciever in the trailer

**+ Fifth wheel lock**
Remotely controlled

**+ Shock absorbers**
Six on first axle, four on second and fourth axles, and two on third axle

**+ Cabin**
With seats and a working steering wheel

**+ Trailer catch guides**
Helps to connect the trailer remotely

**+ Door**
Can be opened manually

**+ Front axles**
Suspended, driven, and steered

**+ Rear axles**
Suspended and driven

## CHALLENGES

The main challenges, aside from writing a custom control program for the NXT unit, were structural. The truck needed a body frame and suspension capable of supporting its heavy, protruding cabin, and the trailer had to carry at least 5.5 lb (2.5 kg) of load without bending. It took some trial and error to achieve that without making the model even heavier.

## THE ORIGINAL

The KZKT-7428 Rusich is a Soviet-built 8×8 prime mover designed to tow loads up to 70 tons across any kind of terrain and under any weather conditions. It's powered by a 650 hp engine with a preheating system that allows it to operate in temperatures as low as -58°F (-50°C). With an 8×8 drive, robust drivetrain, and wheels as tall as an average man, this truck will get you almost anywhere.

# MAN HOOKLIFT TRUCK

*Jennifer Clark (2003)*

## SPECIFICATIONS

| | |
|---|---|
| LENGTH | **18.5"** |
| WIDTH | **5.5"** |
| HEIGHT | **6.7"** |
| PIECES | **~1,800** |

## ABOUT THE MODEL

This model was created with the help of a real hooklift truck driver, who demonstrated all the technical details and workings of the real vehicle. The completed model includes suspension on two rear axles, one of which is driven; steering on two front axles; an elevated container bed; and a pneumatic container catch. All functions are controlled remotely via a third-party radio control system. The body re-creates the livery of the Whiteinch Demolition company, and a complex set of custom stickers is based on the real Whiteinch truck.

## CHALLENGES

Other than fitting multiple electric and pneumatic functions into a rather small chassis, there were many challenges because the model was made before the Power Functions era. Using the weak 9V motors and integrating a third-party control system was difficult, but the biggest challenge was lifting the container bed. Since pneumatics were too weak and linear actuators were not available at the time, the bed is lifted by a custom-built actuator made from studless beams and extended by pulling a string. Simple as it may seem, the resulting mechanism's load capacity is an impressive 2.75 lb (1.25 kg).

### THE ORIGINAL

Hooklift trucks are used to transport containers of various lengths, and with loads up to 30 tons, their arm allows them to act as dump trucks when needed. Hooklift trucks are popular thanks to their ability to drop and pick up loads quickly and accurately even in tight spaces, without the need for the driver to leave the cabin.

# MAN TGS 6×4 CEMENT TRUCK

*Lasse Deleuran (2013)*

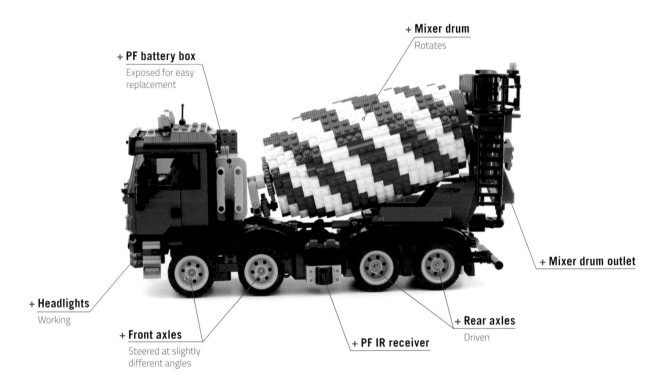

**+ Mixer drum**
Rotates

**+ PF battery box**
Exposed for easy
replacement

**+ Mixer drum outlet**

**+ Headlights**
Working

**+ Rear axles**
Driven

**+ Front axles**
Steered at slightly
different angles

**+ PF IR receiver**

## ABOUT THE MODEL

This MAN TGS cement truck model is built
at 1:25 scale, a classic scale for the official
LEGO Model Team line. The truck's functions
include motorized propulsion and drum rotation, return-to-center steering on its two front
axles, and working headlights. The model is a
great example of functionality, sturdiness, and
an authentic look at a small scale.

## SPECIFICATIONS

| | |
|---|---|
| LENGTH | **14.3"** |
| WIDTH | **5.3"** |
| HEIGHT | **7.8"** |
| PIECES | **1,519** |

### CHALLENGES

Fitting so many functions into a model this small created many challenges; the biggest was creating a steering mechanism that turns the two front axles at slightly different angles. The steering is controlled by a vertically oriented PF Servo motor via a complex system of Technic beams. In addition, the mixer drum was carefully sculpted around an inner structure that can carry cargo, and it is coupled with the propulsion motor.

### THE ORIGINAL

The TGS is the toughest truck produced by the German MAN company. It's built on a robust chassis that can handle both high payloads and off-road conditions. With a high-powered diesel engine, compact layout, and efficiency-oriented overall design, the MAN TGS is an excellent example of a modern European truck.

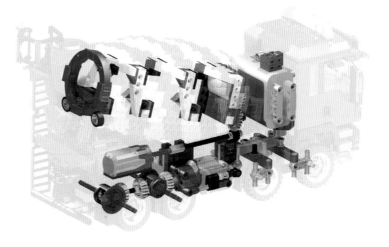

The model's drum is functional—
it can hold LEGO soccer balls and
empty them from its rear chute.
Because of this capability, the truck
was used at the 2011 LEGO World
Fair in Copenhagen as part of the
Great Ball Contraption (GBC) display,
which was a giant collaborative
Rube Goldberg machine used to
transport LEGO soccer balls in the
most spectacular way possible.

# PETERBILT 379 FLATTOP

**ZED (2011)**

### SPECIFICATIONS

| LENGTH | **13.3"** |
| WIDTH | **4.4"** |
| HEIGHT | **6.3"** |

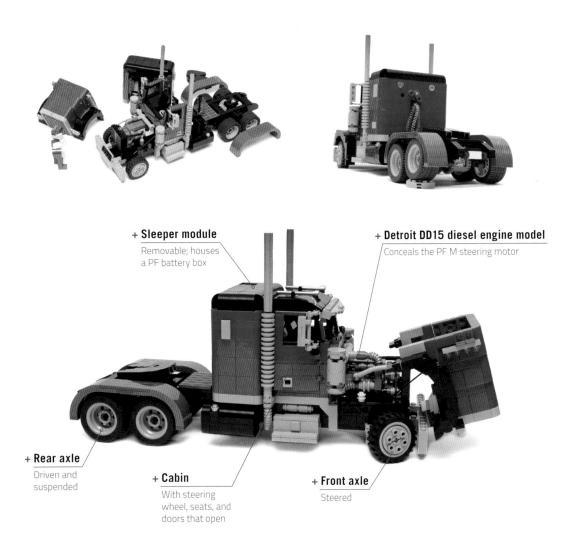

**+ Sleeper module**
Removable; houses
a PF battery box

**+ Detroit DD15 diesel engine model**
Conceals the PF M steering motor

**+ Rear axle**
Driven and
suspended

**+ Cabin**
With steering
wheel, seats, and
doors that open

**+ Front axle**
Steered

### ABOUT THE MODEL

Although this 1:22 model of the classic Peterbilt 379 truck seems to be built for looks, it
was actually created for racing. Designed for
minimum weight and maximum agility, it's
propelled by a PF XL motor; it's steered by a
PF M motor, which is revealed by opening the
hood; and it carries a full-size battery box in
its inconspicuous sleeper module. The steering motor is incorporated in a model of the real
engine, which includes a wedge belt and radiator, and the cabin has a basic interior and doors
that open.

The two rear axles are both driven and
suspended on small rubber axle joiners. This
model's beautiful exterior with all the mechanisms skillfully concealed is topped only by
its performance.

### CHALLENGES

Building a mean racing machine that looks
nothing like Technic was tough, to say the least.
Concealing all the electric elements including
the huge PF XL motor and a battery box that is
almost as wide as the model itself took a great
deal of ingenuity. The sleeper module is removable for battery replacement!

### THE ORIGINAL

Just like many other Peterbilt trucks, the classic 379 is built to suit the needs of an individual client. The chassis length, engine, size
of the sleeper module, and even the livery
can be chosen by a buyer. This makes many
379s look unique, and it was one of these custom orders—a flattop with a short sleeping
module—that inspired this model.

# SCANIA R 4×2 HIGHLINE

**Lasse Deleuran (2012)**

## SPECIFICATIONS

| | |
|---|---|
| LENGTH | **10.3"** |
| WIDTH | **5.25"** |
| HEIGHT | **6.5"** |
| PIECES | **873** |

## ABOUT THE MODEL

This 1:25-scale model of the Scania R was built with emphasis on authenticity, sturdiness, and functionality. In addition to the steering (which uses Ackermann steering geometry) and drivetrain, the working functions include a fifth wheel's coupling mechanism and a trailer with outriggers that are deployed automatically after reversing, which allows for remotely coupling and decoupling the trailer.

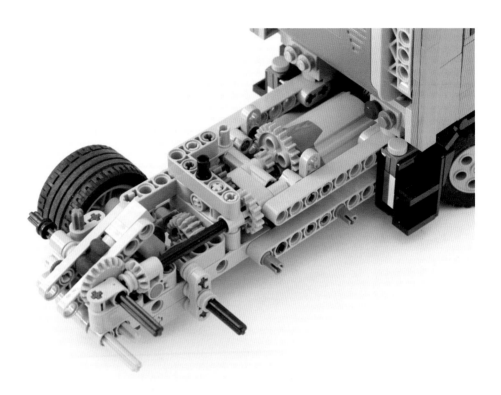

## CHALLENGES

There were two main challenges in this build. The first was creating the rubber band–loaded coupling mechanism that engages as the truck backs up into the trailer and then disengages at a specific point when reversing. The second challenge was the trailer's outrigger mechanism, which is also rubber band–loaded and has a similar working principle. Fitting these elements within a model with working remote steering and propulsion was no small feat.

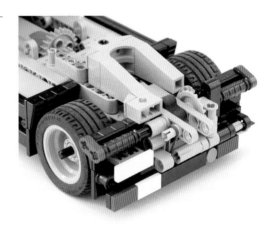

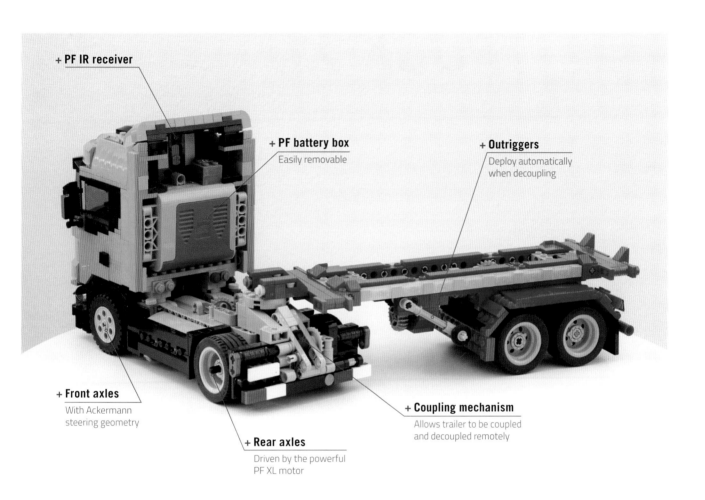

+ **PF IR receiver**

+ **PF battery box**
Easily removable

+ **Outriggers**
Deploy automatically
when decoupling

+ **Front axles**
With Ackermann
steering geometry

+ **Coupling mechanism**
Allows trailer to be coupled
and decoupled remotely

+ **Rear axles**
Driven by the powerful
PF XL motor

## THE ORIGINAL

Scania R is a flagship model from the renowned Swedish Scania company. Available with four cab configurations and a number of engines (including a 730 hp option), it holds the title of the world's most powerful production truck. This 2010 "International Truck of the Year" is considered one of Europe's most beautiful trucks, which makes it a popular choice for LEGO builders.

# TOW TRUCK XL

*Dikkie Klijn (2012)*

### ABOUT THE MODEL

This model of a US-style wrecker features a full suspension with two driven and two steered axles, a rotating extended crane with a double motorized winch, pneumatic outriggers, a fully functional wheel lift, and a full set of working lights complete with a flashing light bar. The entire package is enclosed in a body with plenty of details, custom stickers, opening doors and hood, and lockers in the back that house manual switches controlling some of the functions.

### CHALLENGES

Other than the usual challenges for a model of this size, such as building a sufficiently rigid chassis and concealing all the mechanical parts, the primary challenge was the boom. It had to have an extended section, a reasonable load capacity, and a completely smooth exterior.

## SPECIFICATIONS

| | |
|---|---|
| LENGTH | **32.7"** |
| WIDTH | **7.6"** |
| HEIGHT | **10.2"** |
| PIECES | **4,200** |

### THE ORIGINAL

This is a generic model of a rotator tow truck, with a long boom that can be extended and rotated, plus a wheel lift. The boom is helpful when recovering a vehicle from a ditch or other place that is difficult to get to, and it can be used to return the vehicle to an upright position.

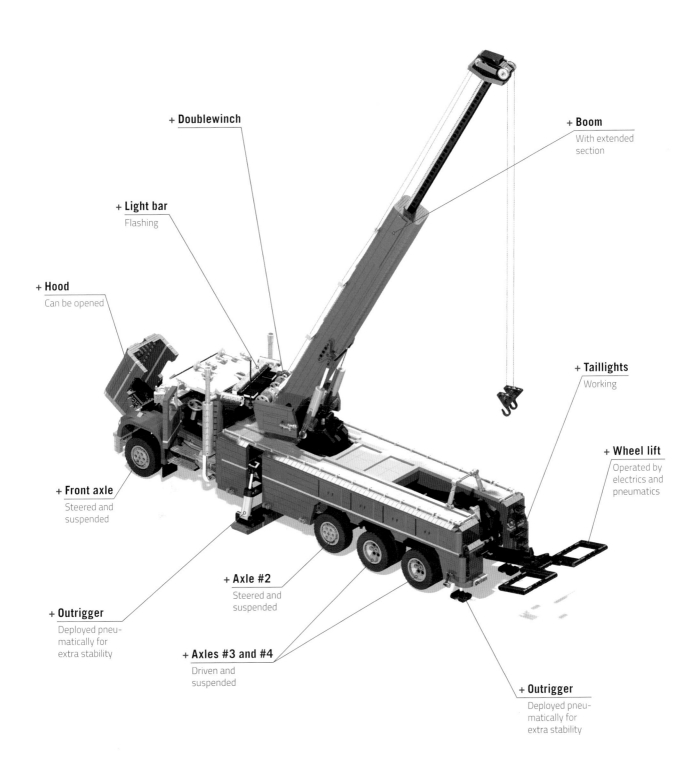

+ **Doublewinch**

+ **Boom**
With extended
section

+ **Light bar**
Flashing

+ **Hood**
Can be opened

+ **Taillights**
Working

+ **Wheel lift**
Operated by
electrics and
pneumatics

+ **Front axle**
Steered and
suspended

+ **Axle #2**
Steered and
suspended

+ **Outrigger**
Deployed pneu-
matically for
extra stability

+ **Axles #3 and #4**
Driven and
suspended

+ **Outrigger**
Deployed pneu-
matically for
extra stability

# 10

# *WATERCRAFT*

# OFFSHORE SUPPORT SHIP

*Efferman (2014)*

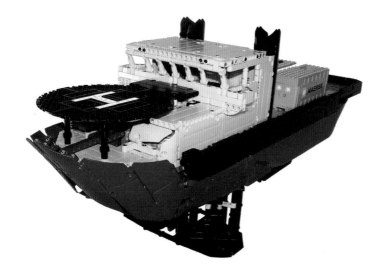

*The cycloidal drive can handle pretty massive vessels; two Voith Schneider propellers acted as auxiliary bow thrusters on the 1938 unfinished German aircraft carrier* **Graf Zeppelin**.

### ABOUT THE MODEL

This model, fully remote controlled and able to operate on water, was built around a unique propulsion system, namely, the Voith Schneider propeller. Sometimes affectionately called an **eggbeater**, it consists of a pair of special turbines that can act as a propeller and a rudder at the same time.

Re-creating the propeller system with LEGO pieces produced quite a large mechanism and required an adequately sized ship to carry it. The Offshore Support Ship weighs 6.5 lb (3 kg) and impresses not only with its ability to float while remaining stable in the water but also with all the advantages of its propulsion system.

It's kept afloat by two unitary 52×12 LEGO hulls under the stern, and three LEGO boxes rest under the bow. The Maersk containers on the rear deck house heavy elements including two PF batteries and PF IR receivers, balancing out the superstructure on the bow. The circular platform near the boat's bow is a helipad.

### CHALLENGES

The two primary challenges were making this complex model float and fitting the propulsion system inside. The two four-bladed propellers are located side-by-side near the bow and driven by a single motor, while another motor controls the pitch of their blades and effectively the direction of the entire ship. The fin under the sterns acts as a support, helping the model rest on flat surfaces when not cruising.

## SPECIFICATIONS

| | |
|---|---|
| LENGTH | **28.7"** |
| WIDTH | **8.5"** |
| HEIGHT | **13.2"** |
| PIECES | **~1,500** |

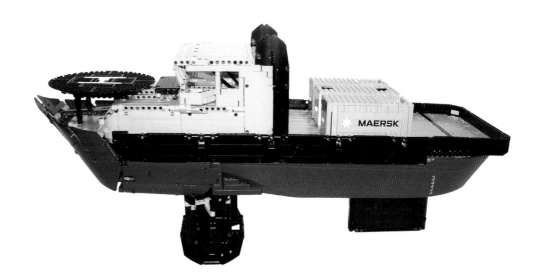

## SIDE VIEW

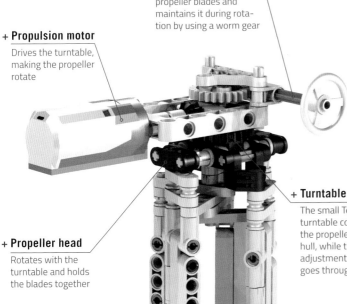

**+ Pitch adjustment axle**
Controls the pitch of all propeller blades and maintains it during rotation by using a worm gear

**+ Propulsion motor**
Drives the turntable, making the propeller rotate

**+ Propeller head**
Rotates with the turntable and holds the blades together

**+ Blade**
Its pitch changes continuously during the rotation, generating thrust in the desired direction.

**+ Turntable**
The small Technic turntable connects the propeller to the hull, while the pitch adjustment axle goes through it.

## THE ORIGINAL

The design of the Voith Schneider propellers originates from a hydroelectric turbine with vertical blades, whose pitch could direct or even reverse the flow of water through it. This creates a so-called cycloidal drive, with two propellers per vessel capable of instantly moving or turning it in any direction. This feature is particularly useful in tugs, ferries, and minesweepers, which rely on their maneuverability. First tested in 1928, the Voith Schneider propellers were quickly introduced in German and then British ships. They have been used by the US Navy as recently as the 1990s.

## BOTTOM VIEW

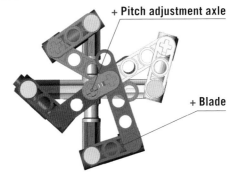

**+ Pitch adjustment axle**

**+ Blade**

# STAN TUG 4011 SL GABON

*Edwin Korstanje (2013)*

## SPECIFICATIONS

| | |
|---|---|
| LENGTH | **55.2"** |
| WIDTH | **15.4"** |
| HEIGHT | **31.5"** |
| PIECES | **~38,000** |

## ABOUT THE MODEL

This 1:30 model of the Stan Tug 4011 SL Gabon tugboat is just the thing to build when you have 90 pounds of LEGO pieces lying around. The hull is built using Technic beams, which make it robust enough for an average person to stand on it. The deck and the superstructure use Technic bricks, which are expertly integrated with the beam-built wheelhouse.

The model, built on commission for the company that owns the original vessel, features two cranes, a motorized winch, working anchors, a pontoon, and 40 LEGO LEDs that bring it to life. As a token of appreciation, the builder was sent on a trip to Africa to see the real Stan Tug tugboat in action.

### CHALLENGES

Other than the sheer size and weight of the model, the main difficulty was combining LEGO Technic beams and LEGO bricks to create a model that is detailed yet sturdy enough to be transported in one piece. The attention to detail included using custom-chromed and copper-plated pieces.

### THE ORIGINAL

Built at the Gelati shipyard in Romania, the Stan Tug 4011 is an all-round offshore tugboat owned by Smit Lamnalco and operating from Gabon, Africa. More than 130 feet (40 meters) long and propelled by two V16 Caterpillar marine engines that develop a total brake power of 5,000 hp, the Stan Tug can function as a rescue and salvage ship, a firefighting vessel, and, of course, a tugboat. It has a top speed of 13.2 knots and a massive bollard pull of almost 75 tons.

# WATER STRIDER

*slfroden (2012)*

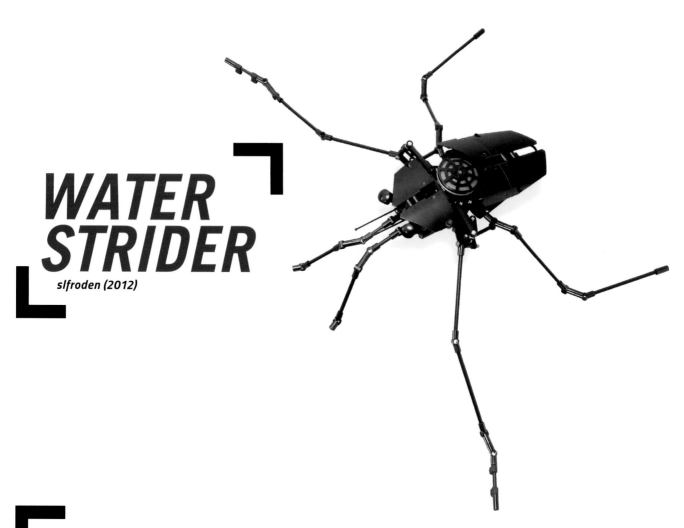

## ABOUT THE MODEL

This Water Strider model is built around a single PF M motor that drives a complex linkage that moves its middle and hind legs in sequence. The finished creation not only looks and moves like a real water strider but—with the addition of four ping-pong balls as flotation devices—can actually walk on water!

## CHALLENGES

The main difficulty was creating a compact, single-motor mechanism that would give the legs the correct range of motion.

Tests on water were quite risky because of the PF M motor's location at the bottom of the thorax. The strider works fine, but it still sank when it hit an obstacle and had to be retrieved from the bottom of a pond.

## THE ORIGINAL

Water striders use their long legs to spread their weight so that the water's surface tension keeps them on top of the water. They steer with their hind legs and propel themselves with the paddle-like movements of their middle legs; they are capable of moving at speeds above 3.28 feet (1 meter) per second. Their forelegs are used for hunting by sensing ripples on the water's surface. Water striders grow up to 1.4 inches (3.5 cm) long and live up to a year. They often fall prey to fish and birds, while their own diet includes, among other things, spiders.

## SPECIFICATIONS

| | |
|---|---|
| LENGTH | **14.4"** |
| WIDTH | **24.7"** |
| HEIGHT | **5.6"** |
| PIECES | **277** |

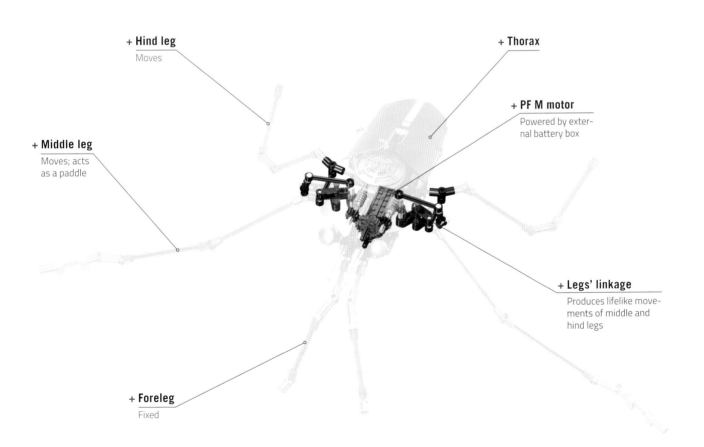

**+ Hind leg**
Moves

**+ Thorax**

**+ PF M motor**
Powered by external battery box

**+ Middle leg**
Moves; acts as a paddle

**+ Legs' linkage**
Produces lifelike movements of middle and hind legs

**+ Foreleg**
Fixed

# BUILDERS

**Ingmar "2LegoOrNot2Lego" Spijkhoven** (The Netherlands)
*http://ingmarspijkhoven.blogspot.nl/*
*http://youtube.com/2LegoOrNot2Lego/*
Models: American Truck, Kenworth 953 Oilfield Truck

**Ingmar "2LegoOrNot2Lego" Spijkhoven** is a full-time freelance builder, one of the lucky few not distracted by a day job. He specializes in 1:17.5-scale models of trucks with trailers and construction equipment. He's renowned for the incredible amount of work he puts into his creations. His models are owned and displayed by Bell Equipment and Elphinstone Construction and by private collectors in Australia, Europe, and the United States—yet he doesn't have a single LEGO set in his house.

**Andrea "6Lovers" Grazi** (Italy)
*http://www.brickshelf.com/cgi-bin/gallery.cgi?m=grazi*
Model: Bigfoot II

Born in the city of Modena, home to most of the Italian sports car makers, and working as an engineer designing mechanical parts for construction equipment, **Andrea "6Lovers" Grazi** had practically no choice but to start building with LEGO Technic. He has a long history of building impressive models with unique technical solutions, even working in the challenging era before Power Functions. He is known for his off-road cars with complex suspension systems and his trucks with incredibly detailed exteriors.

**Amida** (South Korea)
*http://amida.kr/*
Model: Pegasus Automaton

**Amida** has years of professional experience building LEGO dioramas for displays in LEGO shops. He has also translated several LEGO books on the Korean market. As a builder, he's fond of experimenting—some of his creations are built primarily with pneumatic tubes or with oddly shaped Bionicle pieces. Other models surprise with their functionality, such as his caliper for measuring the golden ratio.

**Marat Andreev** (Russia)
*http://youtube.com/gothmog6565/*
Models: Kamaz Dakar Rally Truck, Kawasaki Vulcan 800

A US-based research assistant in physics, **Marat Andreev** finds playing with LEGO Technic relaxing. His strengths are cars and motorcycles with bodies made from classic bricks that are supported by Technic-made chassis. He is known for his success with challenging themes, such as small-scale off-road vehicles, and with embedding Technic functions into motorcycle models. He's also highly skilled in creating multicolor liveries for his models and making custom stickers.

**Barry "Barman76" Bosman** (The Netherlands)
*https://www.flickr.com/photos/50191917@N06/*
*http://youtube.com/barebos/*
Model: SA-2 Samson Battle Helicopter

An engineer by trade, **Barry "Barman76" Bosman** has a reputation for being a pillar of the Technic community. He's been creating awesome models since forever, he freely shares his building ideas and concepts of LEGO pieces not in production, he helps to organize events for the Dutch LEGO community, and he lends a hand when other builders need advice or top-quality photos of their models. As a builder, he's best known for his enormous model of a V8 engine with many moving parts, including 32 valves. His ultimate ambition is to build a working robot that can self-transform.

**Bruno "brunojj1" Jenson** (Germany)
*http://youtube.com/brunojj1/*
Model: Ferrari 458 Spider

**Bruno "brunojj1" Jenson** is an industrial manager who is equally skilled with studless and studfull building styles, as proved by his many models, which include tanks, supercars, and even a helicopter. He appreciates being able to inspire other builders, which is why his latest models come with building instructions.

**Jennifer Clark** (Scotland)
*http://www.genuinemodels.com/*
*http://youtube.com/jenniferclarklego/*
Models: Demag AC50-1, JCB JS220, MAN Hooklift Truck

**Jennifer Clark** created a number of Technic models in the early 2000s that were simply beyond anything anyone previously built. Resulting from extensive research, the models have extremely detailed bodies covered with custom stickers, with functionality powered by third-party mechanical and remote-control elements. Jennifer's amazing models have attracted countless builders to LEGO Technic.

While Jennifer has stopped building and has become a successful professional bassist, her models continue to inspire people all over the world. Even today, with the growing array of Power Functions elements, it's challenging to build something that works and looks half as good as her models. She modestly explains that she just wanted to make accurate working models that looked good.

**Paul James "Crowkillers" Boratko III** (United States)
*http://www.crowkillers.com/*
*http://youtube.com/crowkillers/*
Model: Lamborghini Gallardo, Monster Truck, Muscle Car, Vampire GT

**Paul James "Crowkillers" Boratko III** is a longtime specialist in Technic supercars, which should come as no surprise, considering that he's a real-life automotive technician. His models, which include both real and fictional cars, always look like they could be official LEGO sets and impress with their looks, suspension systems, and transmissions. He is a pioneer in custom chroming LEGO pieces, and he has auctioned a number of his models for charity. In his spare time, he occasionally writes articles on cars and builds engine replicas and mechanical modules on request, yet he's so busy that he built only three official LEGO sets in the last 20 years!

**Lasse Deleuran** (Denmark)
*http://c-mt.dk/*
*http://youtube.com/LasseDeleuran/*
Models: MAN TGS 6×4 Cement Truck,
Scania R 4×2 Highline

Born near Billund, the hometown of the LEGO Group, **Lasse Deleuran** had pretty much no choice but to start building. He became fascinated with LEGO trucks at the age of 4, and they have been his specialty ever since. He likes to build at small scale, which is not only unusual but also quite challenging when you want complex functionality in a model. Over the years, Lasse has become renowned for the "there's-no-way-you-could-fit-motors-in-there!" style.

**Han "Designer-Han" Crielaard** (The Netherlands)
*http://designer-han.nl/*
*http://youtube.com/wuppiesoft/*
Models: Articulated Hauler 6×6, Dump Truck 10×4,
Prinoth Leitwolf

**Han "Designer-Han" Crielaard** is well known among AFOLs for impressive modifications of the official Technic sets, as well as for his own official-looking models, especially trucks. Attracted to building as an adult by the famous Mobile Crane set (#8421), Han likes to take his time, creating a few truly outstanding models per year, complete with building instructions.

**Dirk "Dikkie" Klijn** (The Netherlands)
*http://www.dirkklijn.com/*
*http://youtube.com/DikkieKlijn/*
Models: McLaren MP4-12C, Tow Truck XL

One of the youngest members of the LEGO Technic community, **Dirk "Dikkie" Klijn** is already quite an accomplished builder. He has demonstrated his skill and versatility with his many vehicular models, including trucks, supercars, and snow groomers. He won the official annual LEGO Technic contest at the age of 14. The ever-increasing complexity and superb looks of his creations leave no doubt that Dirk is a serious rising talent.

**Maciej "Dmac" Szymański** (Poland)
*https://www.flickr.com/photos/dmaclego/*
*http://youtube.com/dmaclego/*
Model: DT-75 Tractor

Making his living as a book translator, **Maciej "Dmac" Szymański** is in no rush to build. He rarely reveals more than one model per year, but when he does, jaws hit the floor. While fond of construction equipment and World War II–era tanks, he also has the rare distinction of being the inspiration behind an official LEGO set. In an unprecedented move by the LEGO Group, the massive and impressive Imperial Shuttle set (#10212) was closely based on his model.

**Antti "Drakmin" Hakala** (Finland)
*https://www.flickr.com/photos/drakmin/*
*http://youtube.com/drakmin/*
Models: T-47 Airspeeder "Rebel Snowspeeder," StarCraft
Siege Tank

A specialist in architecture and 3D modeling, **Antti "Drakmin" Hakala** builds LEGO models that look realistic despite being built with beams and panels. He is known as one of the most talented photographers among AFOLs, and he proved his worth as a Technic builder by making something many believed impossible—a functional StarCraft Siege Tank model. He likes to put a lot of work into both a model and its presentation, revealing only about one model per year. His ambition is to build a 4-foot (1.2-meter) long X-Wing model.

**Michael "Efferman" Wirth** (Germany)
*https://www.flickr.com/photos/57623735@N08/*
*http://youtube.com/3fferman/*
Model: Offshore Support Ship

With 28 years of experience in building with LEGO, **Michael "Efferman" Wirth** is a renowned and versatile builder. Trucks, cars, boats, walking machines, tracked vehicles—he has built them all, modding official Technic sets as well. He's known for re-creating unusual machines, such as fire engines or suction excavators, but that's not so strange; he builds suction excavators in real life.

**Michał "Eric Trax" Skorupka** (Poland)
*https://www.facebook.com/pages/Eric-Trax/*
*167670136660216*
*http://youtube.com/erictrax/*
Models: Holmer Terra Dos T3, Ursus C-360-3P

**Michał "Eric Trax" Skorupka**'s specialty is building realistic agricultural machines, but he has plenty of experience with military and off-road vehicles, too. He focuses on combining Model Team–style bodies with complex Technic innards and likes to push the boundaries of size and scale, as he did with his enormous, scale-tipping Holmer Terra sugar beet harvester. With just a few agricultural models, he has already managed to draw a good deal of attention from agriculture trade magazines, which attests to the quality of his work.

**Gyuta** (South Korea)
*http://www.mocpages.com/home.php/7327*
*http://youtube.com/Gyuta97/*
Model: White Tiger T1H1

Making his living as an automotive engineer, **Gyuta** is more than qualified to build Technic models. Reintroduced to building with LEGO by the MINDSTORMS® NXT set, Gyuta puts the functionality of his models above everything else. Specializing in machinery and robotics, he enjoys competing in building contests and using LEGO pieces for charity fundraising.

**Jarek "Jerac" Książczyk** (Poland)
*http://scharnvirk.deviantart.com/*
*http://www.flickr.com/photos/jerac/*
*http://youtube.com/Scharnvirk/*
Model: Land Raider

**Jarek "Jerac" Książczyk**, also known as Scharnvirk, is a software developer by day and a LEGO magician by night. With many hundreds of models under his belt, he is widely known for his series of *Warhammer 40,000* and *StarCraft* models, and he has participated in several official LEGO workshops. His incredibly detailed models range from tiny creatures made with a handful of pieces to a 6.6-foot (2-meter) long, 110 lb (50 kg) Star Destroyer.

**Brian "Klaupacius" Cooper** (United States)
*http://teknomeka.com/*
Model: Tekonomecha

**Brian "Klaupacius" Cooper** divides his time between working as a 3D software engineer and building Japanese-style giant robots. He typically painstakingly builds one large model per year, but he also enjoys creating smaller, motorized battlebots. His fascination with LEGO started with the Moon Landing set (#367) way back in 1975.

**Arjan "Konajra" Oude Kotte** (The Netherlands)
*http://www.konajra.com/*
*http://youtube.com/konajra/*
Models: Caterpillar 7495 HF, Sandvik PF300

**Arjan "Konajra" Oude Kotte** specializes in models of ships and mining equipment at minifig scale. Bicycle mechanic by day, he relies heavily on computer-aided design for LEGO work and likes to take the time to cram as many details as possible into his creations. He often gets commissions from various marine and mining companies; he built a 10.6-foot (3.2-meter) long ship for LEGO World Copenhagen 2015.

**Jurgen Krooshoop** (The Netherlands)
*http://www.jurgenstechniccorner.com/*
*http://youtube.com/Rhodeslover1/*
Models: Koenigsegg CCX, Zorex Excavator

**Jurgen Krooshoop** specializes in supercars and construction equipment. With more than 30 years of building experience, he focuses on creating high-quality building instructions for his own models as well as those by other renowned builders. He's also an avid modder of official Technic sets and prides himself on embedding nine separate remote-controlled functions into one model.

**Francisco Hartley Lyon** (Chile)
*http://www.mocpages.com/home.php/74438*
*http://youtube.com/channel/UCRqMnbkEpe4buYvoDxjigoA/*
Model: Lamborghini Aventador

A professional architect, **Francisco Hartley Lyon** specializes in Technic supercars. He has an excellent eye for detail and proportion in his models, which allows him to capture the look of famous supercars spot on. He's also a talented photographer, putting plenty of effort into presenting his models. He's one of the few builders who prefers to build his supercars without a motor, which allows him to focus more on their aesthetics and mechanical side.

**Madoka "Madoca" Arai** (Japan)
*https://plus.google.com/photos/117021167471864977943/albums?pageid=101385834756929687130*
*http://youtube.com/madoca1977/*
Model: AWD SUV Mk2

**Madoka "Madoca" Arai** is a Technic builder to the core, specializing in incredibly compact cars with beautiful, sleek bodies. His ability to pack an unbelievable amount of function into a tiny space and his skill for getting the look of a real vehicle right with just a handful of pieces have gained him plenty of well-deserved recognition in a short time. He's also a gifted photographer, and he provides free building instructions for each creation. From cars to buggies to trucks, all his models look like they could be official Technic sets—only better.

**Peer "Mahj" Kreuger** (The Netherlands)
*http://vayamenda.com/*
*http://youtube.com/mahj/*
Models: Bridgelayer, Da Vinci Flyer, Stilzkin Indrik, Tachikoma

**Peer "Mahj" Kreuger**'s specialty is finding surprising uses for unusual LEGO pieces. His creations tend to include Bionicle, DUPLO, or even Friends elements and offer a skillful blend of looks and functionality. He has a reputation as a leading Technic video maker, having even developed complex motorized LEGO camera rigs to film his models in motion.

**Maciej "Makorol" Korolik** (Poland)
*http://youtube.com/makoroll/*
Models: Liebherr HS 855 HD, Liebherr LTM 1050-3.1

**Maciej "Makorol" Korolik** is a student of mechanics and machine design, which certainly helps him when working on his remarkable models. A man of many talents, including being a successful musician, Maciej demonstrated his skills with LEGO incredibly early; he was admitted to an adults-only community of Polish builders at the age of 10. Since then, he's been focusing on trucks and construction equipment with Technic innards and flawless Model Team–style exteriors. Unlike many adult builders, he never experienced a "dark age," and the community hopes he never will.

**Marek "M_Longer" Markiewicz** (Poland)
*https://www.flickr.com/photos/m_longer/*
*http://youtube.com/M1longer/*
Models: Kenworth W900L Dump Truck, Liebherr L 580, Liebherr PR 764 Litronic, Sandvik LH 517L

Working as an electrician in an underground mine, **Marek "M_Longer" Markiewicz** has plenty of opportunities to watch mining equipment at work. He is known for models of unique underground vehicles, many of them built at minifig scale and all of them incredibly functional and good looking. His models of Sandvik machines have attracted the manufacturer's attention, as well as the attention of his employer and co-workers. Sometimes called "the long man" because of his 6.5-foot frame, Marek says he received his first LEGO set at age 4 and has kept going ever since.

**Nicolas "Nico71" Lespour** (France)
*http://www.nico71.fr/*
*http://youtube.com/nico71240/*
Models: Braiding Machine, CVT Trophy Truck, Lanz Bulldog Hot Bulb Tractor, Morgan 3 Wheeler

**Nicolas "Nico71" Lespour** is a mechanical engineer who started playing with LEGO Technic in 2008 and went on to become one of the most renowned French builders. Always trying to create something nobody has done before, he has built a number of mechanical clocks and calculators, kinetic sculptures, planimeters, and a working mechanical loom (which got him a job offer from an actual loom manufacturer). He also enjoys building cars, but even his vehicle designs contain marvelous mechanical surprises, like his CVT truck.

**Nathanaël "NKubate" Kuipers** (The Netherlands)
*http://NKubate.com/*
Model: Jeep Hurricane

**Nathanaël "NKubate" Kuipers** is a Dutch design professional who worked for several years as a product developer for the LEGO Group in Denmark. He's the mastermind behind several noteworthy Technic models, including the Snowmobile (#8272), the Cherrypicker (#8292), and the impressive F1 Ferrari Racer (#8674). His other LEGO specialty is alternate builds—that is, creations built from a single set of bricks. He's the author of *The LEGO Build-It Book: Amazing Vehicles* and the second volume, *More Amazing Vehicles* (No Starch Press, 2013), each of which showcases 10 of his creative redesigns of the Super Speedster set (#5867).

**陳彥璋 "Oryx Chen"** (Taiwan)
*http://www.mocpages.com/home.php/96025*
Model: Honda CBR1000RR Repsol

A student of interior design, 陳彥璋 **"Oryx Chen"** has brought the art of creating custom stickers for LEGO models to a whole new level. He's an extremely thorough builder, spending roughly two years on a single creation, occasionally with extra variants. He likes to combine his love for the LEGO Model Team theme with his fondness for working Technic functions, and given the amount of time he spends polishing each model, he's hard to top.

**Luca "RoscoPC" Rusconi** (Italy)
*http://www.roscopc.it/*
Models: Eagle Weslake T1G, McLaren MP4/4

**Luca "RoscoPC" Rusconi** is a true specialist, focusing solely on Formula 1 cars. He has built a dozen models since 2006, picking the most interesting specimens from the 1960s, 1970s, and 1980s and re-creating them with the utmost care, not only including every little detail of their appearance but also creating complex suspension, steering, and drivetrain systems in the process. All his models are built in the same scale, making his stable of race cars extremely impressive when displayed as a whole. He served as a LEGO Ambassador for Italy and his models caught the attention of Clive Chapman, son of the founder of Lotus Cars.

**Senator Chinchilla** (United States)
*http://www.mocpages.com/home.php/16304*
*http://youtube.com/channel/UCj80jQNQUDChPGS38oYwFag/*
Model: Lamborghini Miura Jota

A car lover, **Senator Chinchilla** devotes most of his building activities to models of real vehicles with carefully re-created bodies. Occasionally dabbling in sci-fi vehicles and spaceships, he's a productive builder, with more than 100 creations. He considers the look of a model to be far more important than its functions, and he shows minimal interest in modern beams, ensuring his models have a nostalgic, old-school vibe.

**Fernando "Sheepo" Benavides de Carlos** (Spain)
*http://www.sheepo.es/*
*http://youtube.com/Sheepo86/*
Models: Bugatti Veyron 16.4 Grand Sport, Ford Mustang Shelby GT500, Land Rover Defender 110, Porsche 911 (997) Turbo Cabriolet PDK

**Fernando "Sheepo" Benavides de Carlos** is a prolific builder of large-scale, extremely functional models of the world's most expensive cars. Model after model, he invents and perfects mechanisms such as disc brakes, multispeed remote-controlled transmissions, and automated clutches. He also inspires many fellow builders by creating instructions, and every time you think his latest model can't possibly be topped, he comes up with something new. His cars have been featured in many magazines and on TV shows around the world, and they have won a number of contests including the LEGO Group's 2011 Technic Challenge. And he always seems to have a new model up his sleeve!

**Stephan "slfroden" Froden** (Australia)
*http://www.splat-design.com/*
*http://youtube.com/slfroden/*
Model: Water Strider

For **Stephan "slfroden" Froden**, building with Technic is all about the intriguing mechanisms. That's why many of his creations are subsystems, such as gear-boxes and various couplings and controllers. He especially likes complex linkages that perform complicated movement—and he used such a mechanism to build his award-winning Water Strider. While most builders would be content with this movement alone, he proceeded to make this large, motorized creation stay afloat and move on water, which is a great testament to his ingenuity.

**Pablo "Spiderbrick" Alvarez Espinoza** (Costa Rica)
*http://www.brickshelf.com/cgi-bin/gallery.cgi?m=spiderbrick*
*http://youtube.com/pabloalvarezcr/*
Model: Volkswagen Jetta

**Pablo "Spiderbrick" Alvarez Espinoza** is a computer engineer with a thing for Technic supercars. He has experience re-creating sports cars—but unlike most builders who compete to build models of the most exotic and expensive cars possible, he prefers building everyday cars. Making a regular sedan with all the functionality and building techniques of top LEGO supercars is a refreshing initiative of his that has sparked a lot of interest among other builders.

**Kyle "Thirdwigg" Wigboldy** (United States)
*http://thirdwigg.com/*
*http://youtube.com/thirdwigg/*
Model: Spitfire

**Kyle "Thirdwigg" Wigboldy** has more than 10 years of experience building LEGO models, dabbling in planes, cars, trucks—pretty much anything that moves. He is comfortable with various building techniques and has amazed the Technic community by combining Technic-on-the-inside style with Model-Team-on-the-outside techniques to build large-scale models of historic aircraft.

**Edwin "VFracingteam" Korstanje** (The Netherlands)
*https://www.flickr.com/photos/vfracingteam/*
*http://youtube.com/VFracingteam/*
Model: Stan Tug 4011 SL Gabon

**Edwin "VFracingteam" Korstanje** specializes in boats and trucks, and his trademark technique is placing beams on their sides to avoid showing pin holes. He began building in 2011 and quickly became a widely recognized builder, building mostly on request for various companies around the world. His massive, detail-rich models have been covered by the media in more than 60 countries, and three of the world's leading industrial companies display his models at their headquarters.

**Ignat "ZED" Khliebnikov** (Ukraine)
*http://www.brickshelf.com/cgi-bin/gallery.cgi?m=ZED*
*http://youtube.com/ZEDDoubleBrick/*
Models: Caterpillar D9T, Peterbilt 379 Flattop

A professional developer of educational software and robots, **Ignat "ZED" Khliebnikov** is the originator of 1:22-scale LEGO truck racing and a longtime coach of the Ukrainian team for the World Robot Olympiad. He occasionally finds time to enjoy ordinary Technic—but always with amazing results. He focuses on details and photorealistic appearance.

# CREDITS

All copyrights retained by the individual copyright holders. All photographs provided by the builders of their respective models, except for those listed below.

For a clickable list of these links, visit *http://nostarch.com/techniclinks/*.

### AMERICAN TRUCK
Builder: Ingmar Spijkhoven
Video: *http://youtu.be/_O_vOH3CNm8*
Building Instructions: *http://mocplans.com/designer/ingmar-spijkhoven/*
More Info: *http://mocpages.com/moc.php/163114*

### ARTICULATED HAULER 6×6
Builder: Han Crielaard
Image Credits: Eric Albrecht
Video: *http://youtu.be/PRPho2yLY_Y*
Building Instructions: *http://designer-han.nl/lego/*

### AWD SUV MK2
Builder: Madoka Arai
Video: *http://youtu.be/Y-JJgfLKBLM*
Building Instructions: *https://plus.google.com/u/0/photos/117021167471864977943/albums/5901503914497924817*

### THE BAT
Builder: Paweł Kmieć
Video: *http://youtu.be/xHLiAsVit_k*
More Info: *http://sariel.pl/2013/08/the-bat/*

### BIGFOOT II
Builder: Andrea Grazi
Building Instructions: *http://www.brickshelf.com/cgi-bin/gallery.cgi?f=31042*

### BRAIDING MACHINE
Builder: Nicolas Lespour
Video: *http://youtu.be/I9B1hqcAt1s*
Building Instructions: *http://www.nico71.fr/braiding-machine-makes-wristband/*

### BRIDGELAYER
Builder: Peer Kreuger
Video: *http://youtu.be/Mcr5VtM5aEw*

### BUGATTI VEYRON 16.4 GRAND SPORT
Builder: Fernando Benavides de Carlos
Video: *http://youtu.be/jHWDSnWk2jU*
More Info: *http://www.sheepo.es/2011/03/bugatti-veyron-164-grandsport.html*

### CATERPILLAR 7495 HF
Builder: Arjan Oude Kotte
More Info: *http://www.konajra.com/#!latest-projects*

### CATERPILLAR D9T
Builder: Ignat Khliebnikov
Video: *http://youtu.be/FH1bTvQukpA*
Building Instructions: *http://www.brickshelf.com/cgi-bin/gallery.cgi?f=404695*

### CVT TROPHY TRUCK
Builder: Nicolas Lespour
Video: *http://youtu.be/VQC62bwnvE4*
Building Instructions: *http://www.nico71.fr/trophy-truck-with-continuously-variable-transmission/*

### DA VINCI FLYER
Builder: Peer Kreuger
Video link: *http://youtu.be/-g6U0bWXjto*

### DEMAG AC50-1
Builder: Jennifer Clark
Video link: *http://youtu.be/F_tMgkuqiS4*
Building Instructions: *http://mocplans.com/demag-ac50-all-terrain-crane.html*
More Info: *http://genuinemodels.com/demag_crane.htm*

### DT-75 TRACTOR
Builder: Maciej Szymański

### DUMP TRUCK 10×4
Builder: Han Crielaard
Image Credits: Eric Albrecht
Video: *http://youtu.be/spdcgaesIds*
Building Instructions: *http://designer-han.nl/lego/*

### EAGLE WESLAKE T1G
Builder: Luca Rusconi
Photos Credits: Marco Angeretti
Building Instructions: *http://mocplans.com/designer/roscopc*
More Info: *http://roscopc.it/#/2*

### FERRARI 458 SPIDER
Builder: Bruno Jenson
Video: *http://youtu.be/33ZVMv5KPnk*
Building Instructions: *http://mocplans.com/red-spider.html*

### FORD MUSTANG SHELBY GT500
Builder: Fernando Benavides de Carlos
Video: *http://youtu.be/fMKe89zcvHU*
Building Instructions: *http://www.sheepo.es/2014/01/ford-mustang-shelby-gt500-instructions.html*
More Info: *http://www.sheepo.es/2013/10/ford-mustang-shelby-gt500-14.html*

### HOLMER TERRA DOS T3
Builder: Michał Skorupka
Video: *http://youtu.be/4gbtwEP6d3E*
More Info: *http://www.eurobricks.com/forum/index.php?showtopic=89949*

### HONDA CBR1000RR REPSOL
Builder: 陳彥璋 [Oryx Chen]
More Info: *https://ideas.lego.com/projects/30154*

### HUMMER H1 WAGON
Builder: Paweł Kmieć
Video: *http://youtu.be/P48MXed3EOw*
Building Instructions: *http://sariel.pl/2014/06/hummer/*

### JEEP HURRICANE
Builder: Nathanaël Kuipers
Video: *http://youtu.be/-sAHW-CR3DM*
Building Instructions: *http://nkubate.com/index.php?option=com_k2&view=item&id=29:jeep-hurricane&Itemid=566*

### JCB JS220
Builder: Jennifer Clark
Video: *http://youtu.be/MIcgzhivbjQ*
Building Instructions: *http://mocplans.com/js220-excavator.html*
More Info: *http://genuinemodels.com/jcb.htm*

## K2 BLACK PANTHER
Builder: Paweł Kmieć
Video link: *http://youtu.be/c0JpkmKajOs*
More Info: *http://sariel.pl/2013/03/k2-black-panther/*

## KAMAZ DAKAR RALLY TRUCK
Builder: Marat Andreev
Video: *http://youtu.be/znQQMM03cdI*

## KAWASAKI VULCAN 800
Builder: Marat Andreev

## KENWORTH 953 OILFIELD TRUCK
Builder: Ingmar Spijkhoven
Video: *http://youtu.be/NiTeiKLfplk*
Building Instructions: *http://mocpages.com/moc.php/282973*

## KENWORTH W900L DUMP TRUCK
Builder: Marek Markiewicz
Video: *http://youtu.be/nupnwQxKsTU*

## KOENIGSEGG CCX
Builder: Jurgen Krooshoop
Photo Credits: Barry Bosman
Video: *http://youtu.be/HDo2iNS6V2A*
Building Instructions: *http://jurgenstechniccorner.com/instructies.html*
More Info: *http://jurgenstechniccorner.com/koenigsegg.html*

## KZKT-7428 RUSICH
Builder: Paweł Kmieć
Video: *http://youtu.be/ly40f1aQ-sY*
More Info: *http://sariel.pl/2013/05/kzkt-7428-rusich/*

## LAMBORGHINI AVENTADOR
Builder: Francisco Hartley Lyon
Video: *http://youtu.be/_3I1KSZlaUw*
More Info: *http://mocpages.com/moc.php/360466*

## LAMBORGHINI GALLARDO
Builder: Paul Boratko
Building Instructions: *http://crowkillers.com/model.php?model=gallardo*

## LAMBORGHINI MIURA JOTA
Builder: Senator Chinchilla
More Info: *http://mocpages.com/moc.php/377397*

## LAND RAIDER
Builder: Jarek Ksiazcyk
Video: *http://youtu.be/qEx2LQY_1W0*

## LAND ROVER DEFENDER 110
Builder: Fernando Benavides de Carlos
Video: *http://youtu.be/aXW-7DVlyU0*
Building Instructions: *http://www.sheepo.es/2013/02/land-rover-defender-110-instructions.html*
More Info: *http://www.sheepo.es/2012/05/land-rover-defender-110.html*

## LANZ BULLDOG HOT BULB TRACTOR
Builder: Nicolas Lespour
Video: *http://youtu.be/rxphaNrF5pA*
Building Instructions: *http://www.nico71.fr/hot-bulb-pneumatic-tractor/*

## LIEBHERR HS 855 HD
Builder: Maciej Korolik
Video: *http://youtu.be/zwNfCSLac8c*

## LIEBHERR LTM 1050-3.1
Builder: Maciej Korolik
Video: *http://youtu.be/SY-7mQMUkdk*

## LIEBHERR L 580
Builder: Marek Markiewicz
Video: *http://youtu.be/Ln-6z8FWG-4*
Building Instructions: *http://mocplans.com/liebherr-l580-front-loader.html*

## LIEBHERR PR 764 LITRONIC
Builder: Marek Markiewicz
Video: *http://youtu.be/lpjzQztVdZ0*

## LOCKHEED SR-71 BLACKBIRD
Builder: Paweł Kmieć
Video: *http://youtu.be/BxhrOjYO7IY*
More Info: *http://sariel.pl/2013/06/sr-71-blackbird/*

## MAN HOOKLIFT TRUCK
Builder: Jennifer Clark

## MAN TGS 6X4 CEMENT TRUCK
Builder: Lasse Deleuran
Video: *http://youtu.be/HG3OTmA2BjU*
Building Instructions: *http://c-mt.dk/instructions/models_truck-Cement.htm#Cement*

## MCLAREN MP4/4
Builder: Luca Rusconi
Photo Credits: Marco Angeretti
Building Instructions: *http://mocplans.com/designer/roscopc*
More Info: *http://roscopc.it/#/2*

## MCLAREN MP4-12C
Builder: Dirk Klijn
Video: *http://youtu.be/El3p1R00Eu0*
Building Instructions: *http://eurobricks.com/forum/index.php?showtopic=95213*

## MERCEDES-BENZ 540K SPECIAL ROADSTER
Builder: Paweł Kmieć
Video: *http://youtu.be/bcvt0O2BoGc*
Building Instructions: *http://sariel.pl/downloads/*
More Info: *http://sariel.pl/2012/12/mercedes-benz-540k-special-roadster/*

## MONSTER TRUCK
Builder: Paul Boratko
Video: *http://youtu.be/BMUNIy-FmUM*
More Info: *http://crowkillers.com/model.php?model=monster*

## MORGAN 3 WHEELER
Builder: Nicolas Lespour
Video: *http://youtu.be/3MB0G7UpFUw*
Building Instructions: *http://www.nico71.fr/morgan-three-wheeler/*

## MUSCLE CAR
Builder: Paul Boratko
Video: *http://youtu.be/HYz4_SFykjo*
Building Instructions: *http://rebrickable.com/mocs/crowkillers/black-american-muscle-car*
More Info: *http://crowkillers.com/model.php?model=2014musclecar*

## OFFSHORE SUPPORT SHIP
Builder: Michael Wirth
Video: *http://youtu.be/DWRZ2HgYtLA*

## PAGANI ZONDA
Builder: Paweł Kmieć
Video link: *http://youtu.be/DAMb7t1DyZ8*
More Info: *http://sariel.pl/2012/09/pagani-zonda/*

## PEGASUS AUTOMATON
Builder: Amida Na
Video: http://vimeo.com/44369308

## PETERBILT 379 FLATTOP
Builder: Ignat Khliebnikov
Video link: http://youtu.be/CAAegdhNYN8

## PORSCHE 911 (997) TURBO CABRIOLET PDK
Builder: Fernando Benavides de Carlos
Video: http://youtu.be/ZXK5a6IcjLE
More Info: http://www.sheepo.es/2011/05/porsche-911-997-turbo-cabriolet-pdk_04.html

## PRINOTH LEITWOLF
Builder: Han Crielaard
Image Credits: Eric Albrecht
Video: http://youtu.be/2IrVsBoRBCI
Building Instructions: http://designer-han.nl/lego

## SA-2 SAMSON BATTLE HELICOPTER
Builder: Barry Bosman
Video: http://youtu.be/Nnuv-DGsSWE
Building Instructions: http://jurgenstechniccorner.com/instructies.html

## SANDVIK LH 517
Builder: Marek Markiewicz
Video: http://youtu.be/vcj3ld4almM
More Info: http://eurobricks.com/forum/index.php?showtopic=41097

## SANDVIK PF300
Builder: Arjan Oude Kotte
Video: http://youtu.be/C2hGbNnQ0wE
Building Instructions: http://konajra.com/#!latest-projects

## SCANIA R 4×2 HIGHLINE
Builder: Lasse Deleuran
Video: http://youtu.be/JC9Cgz70p30
Building Instructions: http://c-mt.dk/instructions/models_truck-Scania
Highline2.htm#ScaniaHighline2

## SPITFIRE
Builder: Kyle Wigboldy
Video: http://youtu.be/-Jublc1ge1Q
Building Instructions: http://thirdwigg.com/building-instructions/
More Info: http://thirdwigg.com/2012/12/30/spitfire-mk-iia/

## STAN TUG 4011 SL GABON
Builder: Edwin Korstanje
More Info: http://mocpages.com/moc.php/378738

## STARCRAFT SIEGE TANK
Builder: Antti Hakala
Video: http://youtu.be/cpT2eB25NPE
More Info: http://mocpages.com/moc.php/330127

## STILZKIN INDRIK
Builder: Peer Kreuger
Video: http://youtu.be/STzYYgJmsoc
Building Instructions: http://vayamenda.com/

## T-47 AIRSPEEDER "REBEL SNOWSPEEDER"
Builder: Antti Hakala
Building Instructions: http://mocplans.com/designer/rebel-snowspeeder.html
More Info: http://mocpages.com/moc.php/378848

## TACHIKOMA
Builder: Peer Kreuger
Video: http://youtu.be/dargOslomMA
Building Instructions: http://vayamenda.com/

## TEKNOMEKA
Builder: Brian Cooper
Building Instructions: http://teknomeka.com/

## TOW TRUCK XL
Builder: Dirk Klijn
Video: http://youtu.be/cKzjGypsq60
Building Instructions: http://crowkillers.com/instructions.php
More Info: http://www.dirkklijn.com/towtruck/

## TUMBLER
Builder: Paweł Kmieć
Video: http://youtu.be/yUTD9z-aiHU
More Info: http://sariel.pl/2012/12/tumbler/

## URSUS C-360-3P
Builder: Michał Skorupka
Video: http://youtu.be/ty4VpEzXR7o
More Info: http://eurobricks.com/forum/index.php?showtopic=95299

## VAMPIRE GT
Builder: Paul Boratko
Video: http://youtu.be/N4YaR8J14q0
Building Instructions: http://rebrickable.com/sets/crowkillers/vampire-gt-deluxe-black
More Info: http://crowkillers.com/model.php?model=vampire-gt

## VOLKSWAGEN JETTA
Builder: Pablo Alvarez Espinoza
Video: http://youtu.be/IcrCsOHRIJI

## WATER STRIDER
Builder: Stephan Froden
Video: http://youtu.be/ObLStODGZDs
Building Instructions: http://rebrickable.com/mocs/slfroden/technic-water-strider

## WHITE TIGER T1H1
Builder: Gyu-sung Kim
Video: http://youtu.be/OcUhWDkQ6sk
More Info: http://www.mocpages.com/moc.php/244445

## ZOREX EXCAVATOR
Builder: Jurgen Krooshoop
Photo Credits: Barry Bosman
Video: http://youtu.be/pPhrR1EIIR8
Building Instructions: http://jurgenstechniccorner.com/instructies.html
More Info: http://jurgenstechniccorner.com/zorex.html

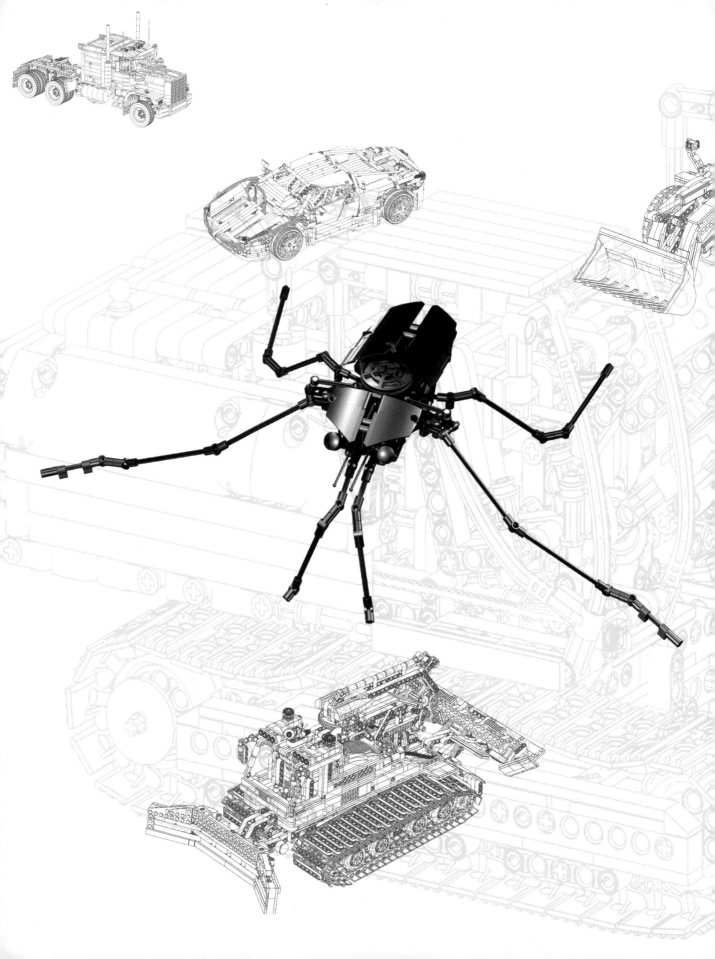